# Practical Electronics
# Handbook

# Practical Electronics Handbook

IAN R. SINCLAIR

Newnes Technical Books

**The Butterworth Group**

| | |
|---|---|
| United Kingdom | **Butterworth & Co. (Publishers) Ltd**<br>London: 88 Kingsway, WC2B 6AB |
| Australia | **Butterworths Pty Ltd**<br>Sydney: 586 Pacific Highway, Chatswood, NSW 2067<br>Also at Melbourne, Brisbane, Adelaide and Perth |
| Canada | **Butterworth & Co. (Canada) Ltd**<br>Toronto: 2265 Midland Avenue, Scarborough,<br>Ontario M1P 4S1 |
| New Zealand | **Butterworths of New Zealand Ltd**<br>Wellington: T & W Young Building, 77—85 Customhouse<br>Quay, 1, CPO Box 472 |
| South Africa | **Butterworth & Co. (South Africa) (Pty) Ltd**<br>Durban: 152—154 Gale Street |
| USA | **Butterworth (Publishers) Inc**<br>Boston: 10 Tower Office Park, Woburn, Mass. 01801 |

First published 1980

© Butterworth & Co. (Publishers) Ltd, 1980

**British Library Cataloguing in Publication Data**

Sinclair, Ian Robertson
  Practical Electronics Handbook
  1. Electronic circuits — Handbooks, manuals, etc.
  I. Title
  621.3815'3'0202     TK7867

ISBN 0-408-00447-9

Typeset by Scribe Design, Gillingham, Kent
Printed in England by Redwood Burn Ltd.,
Trowbridge and Esher

# Preface

Databooks often tend to be simply collections of information, with little or nothing in the way of explanation, and in many cases with so much information that the user has difficulty in selecting what he needs. This book has been designed to include within a reasonable space most of the information which is useful in electronics together with brief explanations which are intended to serve as reminders rather than instruction. The book is not, of course, intended as a form of beginner's guide to the whole of electronics, but the beginner will find here much of interest, as well as a compact reminder of electronic principles and circuits. The constructor of electronic circuits and the service engineer should both find the data in this book of considerable assistance, and the professional design engineer will also find that the items collected here are of frequent use, and would normally only be available in collected form in much larger volumes.

I hope therefore, that this book will become a useful companion to anyone with an interest in electronics, and that the information in the book will be as useful to the reader as it has been to me.

Ian R. Sinclair

# Contents

# Introduction

# Mathematical Conventions

Quantities greater than 100 or less than 0.01 are usually expressed in the *standard form:* $A \times 10^n$ where $A$ is a number less than 10, and $n$ is a whole number. A positive value of $n$ means that the number is greater than unity, a negative value of $n$ means that the number is less than unity. To convert a number into standard form shift the decimal point until a number between 1 and 10 is obtained and count the number of places which have been shifted, which will be the value of $n$. If the decimal point has had to be shifted to the left $n$ is positive, if the decimal point had to be shifted to the right $n$ is negative. For example:

1200 is $1.2 \times 10^3$, but 0.0012 is $1.2 \times 10^{-3}$

To convert numbers back from standard form, shift the decimal point $n$ figures to the right if $n$ is positive and to the left if $n$ is negative. Example:

$5.6 \times 10^{-4} = 0.000\ 56$;    $6.8 \times 10^5 = 680\ 000$

Numbers in standard form can be entered into a scientific calculator by using the key marked Exp or EE (for details, see the manufacturer's instructions).

Numbers in standard form can be used for insertion in formulae, but component values are more conveniently written using the prefixes shown in *Table 0.1.* Prefixes are chosen so that values can be written without using small fractions or large numbers.

Throughout this book, equations have been given in as many forms as are normally needed, so that the reader should not have to transpose

*1*

equations. For example, Ohm's law is given in all three forms, $V = RI$, $R = V/I$, and $I = V/R$. The units which must be used with formulae are also shown, and must be adhered to. For example, the equation: $X = 1/(2\pi fC)$ is used to find the reactance of a capacitor in ohms, using $C$ in farads and $f$ in hertz. If the equation is to be used with values given

**Table 0.1**

| Prefix | Abbreviation | Power of Ten | Decimal multiplier |
|--------|--------------|--------------|--------------------|
| Giga | G | $10^9$ | 1 000 000 000 |
| Mega | M | $10^6$ | 1 000 000 |
| kilo | k | $10^3$ | 1 000 |
| milli | m | $10^{-3}$ | 1/1 000 |
| micro | $\mu$ | $10^{-6}$ | 1/1 000 000 |
| nano | n | $10^{-9}$ | 1/1 000 000 000 |
| pico | p | $10^{-12}$ | 1/1 000 000 000 000 |

*Note that* 1 000 pF = 1 nF; 1 000 nF = 1 $\mu$F and so on
**Examples:**   1 k$\Omega$ = 1 000 $\Omega$ (sometimes written as 1K0, see *Table 1.3*)
1 nF = 0.001 $\mu$F, 1 000 pF or $10^{-9}$ F
4.5 MHz = 4 500 kHz = 4.5 $\times$ $10^6$ Hz

in $\mu$F and kHz, then values such as 0.1 $\mu$F and 15 kHz are entered as 0.1 $\times$ $10^{-6}$ and 15 $\times$ $10^3$. Alternatively, the equation can be written as $X = 1/(2\pi fC)$ M$\Omega$ with $f$ in kHz and $C$ in nF.

In all equations, multiplication may be indicated by use of a dot (A.B) or by close printing ($2\pi fC$). Where brackets ( ) are used in an equation, the quantities within the brackets should be worked out first. For example:

$$2 ( 3 + 5 ) \text{ is } 2 \times 8 = 16 \quad \text{and} \quad 2 + ( 3 \times 5 ) \text{ is } 2 + 15 = 17$$

Transposing, or changing the subject of an equation, is simple provided the essential rule is remembered: an equation is not altered by carrying out identical operations on both sides.

*Example:*   $Y = \dfrac{5aX + b}{C}$ is an equation

If this has to be transposed so that it reads as a formula to find $X$, then the procedure is to change both sides so that $X$ is left isolated. The steps are as follows:

(a)   multiply both sides by $C$   result: $CY = 5aX + b$

(b)   subtract $b$ from both sides   result: $CY - b = 5aX$

(c)   divide both sides by $5a$   result: $\dfrac{CY - b}{5a} = X$

Now the equation reads $X = \dfrac{CY - b}{5a}$ which is the desired transposition.

# Chapter 1

# Passive Components

## Resistors

Resistance, measured in ohms ($\Omega$), is defined as the ratio of voltage (in volts) across a length of material to current (in amperes) through the material. When a graph is drawn of voltage across the material plotted

(a)

(b)

*Figure 1.1. (a) A circuit for checking the behaviour of a resistor. (b) The shape of the graph of voltage plotted against current for an ohmic resistor, using the circuit in (a).*

against current through the material, the value of resistance is represented by the *slope* of the graph. For a material which is kept at a constant temperature, a straight-line graph indicates that the material is *ohmic*, obeying Ohm's law (*Figure 1.1*). Non-ohmic behaviour is represented on such a graph by curved lines or lines which do not pass through the point, called the origin, which represents zero voltage and

*Figure 1.2. Three types of non-ohmic behaviour indicated by graph curves*

zero current. Non-ohmic behaviour can be caused by temperature changes (light bulbs, thermistors), by voltage-generating effects (thermocouples), or by conductivity being affected by voltage (diodes), as in *Figure 1.2*.

Resistance values on components are either colour coded, as noted in *Table 1.2*, or have values printed on using the convention of BS 1852: 1970 (*Table 1.3*).

## *Resistivity*

The resistance of any sample of material is determined by its dimensions and by the value of resistivity of the material. Wire drawn from a single reel will have a resistance value depending on the length cut; for example, a 3 m length will have three times the resistance of a 1 m length. When equal lengths of wire of the same material are compared, the resistance multiplied by the square of the diameter is the same for each. For example, if a given length of a sample wire has a resistance of 12 ohms and its diameter is 0.3 mm, then the length of wire of diameter 0.4 mm, of the same material, will have resistance $R$ such that

$$R \times 0.4^2 = 12 \times 0.3^2 \text{ so that } R = \frac{12 \times 0.3^2}{0.4^2} = \frac{12 \times 0.09}{0.16} = 6.75 \text{ ohms}$$

Resistivity measures the effect which different materials contribute to the resistance of a wire. The resistivity of the material can be found by measuring the resistance $R$ of a sample, multiplying by the area of cross-section, and then dividing by the length of the sample. As a formula, this is

$$\rho = \frac{R.A}{L}$$

where $\rho$ (Greek rho) is resistivity, $R$ is resistance, $A$ is area of cross-section and $L$ is length. When $R$ is in ohms, $A$ in square metres, and $L$

#### Table 1.2.   RESISTOR COLOUR CODE

1st   2nd   3rd      Tolerance
                      (if used)

**Band Coding**
*First Band* — first figure of resistance value (first *significant* figure)
*Second Band* — second figure of resistance value (second *significant* figure)
*Third Band* — number of zeros following second figure (multiplier)
*Tolerance Band* — percentage tolerance of value (5% or 10%). No tolerance band is used if the resistor has 20% tolerance.

**Code Colours** (also used for capacitor values)

| Figure | Colour | |
|--------|--------|--|
| 0 | black | |
| 1 | brown | |
| 2 | red | |
| 3 | orange | |
| 4 | yellow | |
| 5 | green | |
| 6 | blue | |
| 7 | violet | |
| 8 | grey | |
| 9 | white | |
| 0.01 | silver | } used as multipliers (3rd band) only |
| 0.1 | gold | |
| *Tolerance* | | |
| 10% | silver | |
| 5% | gold | |

#### Table 1.3.   RESISTANCE VALUE CODING

Resistance values on components and in component lists are often coded according to BS 1852. In this scheme no decimal points are used, and a value in ohms is indicated by R, kilohms by K (not k), and megohms by M. The letter R, K or M is used in place of the decimal point, with a zero in the leading position if the value is less than 1 ohm.
**Examples**      1K5 = 1.5k or 1500 ohms; 2M2 = 2.2M$\Omega$; 0R5 = 0.5 ohms

in metres, $\rho$ is in ohm-metres. Since most wires are of circular cross-section, $A = \pi r^2$ (or $1/4\pi d^2$ where $d$ is diameter).

Because the resistivities of materials are well known, this formula is much more useful in the form

$$R = \frac{\rho L}{A}$$

with $\rho$ in ohm-metres, $L$ in metres, $A$ in metres$^2$, to find the resistance of a given length of wire of known area. This formula can be rewritten as

$$R = 1.27 \times 10^{-3} \frac{\rho L}{d^2}$$

with $\rho$ in nano-ohm metres, as in *Table 1.4*, $L$ in metres, and $d$ (diameter) in millimetres. *Table 1.4* shows values of resistivities, in nano-ohm

**Table 1.4.  VALUES OF RESISTIVITY AND CONDUCTIVITY**

The values of resistivity are in nano-ohm metres. The values of conductivity are in megasiemens per metres.

|  | Metal | Resistivity | Conductivity |
|---|---|---|---|
| | Aluminium | 27.7 | 37 |
| | Copper | 17 | 58 |
| | Gold | 23 | 43 |
| | Iron | 105 | 9.5 |
| *Pure* | Nickel | 78 | 12.8 |
| *elements* | Platinum | 106 | 9.4 |
| | Silver | 16 | 62.5 |
| | Tin | 115 | 8.7 |
| | Tungsten | 55 | 18.2 |
| | Zinc | 62 | 16 |
| | Carbon-steel (average) | 180 | 5.6 |
| | Brass | 60 | 16.7 |
| | Constantan | 450 | 2.2 |
| | Invar | 100 | 10 |
| | Manganin | 430 | 2.3 |
| *Alloys* | Nichrome | 1105 | 0.9 |
| | Nickel-silver | 272 | 3.7 |
| | Monel metal | 473 | 2.1 |
| | Kovar | 483 | 2.0 |
| | Phosphor-bronze | 93 | 10.7 |
| | 18/8 stainless steel | 897.6 | 1.11 |

metres, for various metals, both elements and alloys. The calculation of resistance by either formula follows the pattern of the example below.

*Example A:*   Find the resistance of 6.5 m of wire, diameter 0.6 mm, if the resistivity value is 430 nano-ohm metres.

Using $R = \dfrac{\rho L}{A}$, $\rho = 430 \times 10^{-9}$, $L = 6.5$ m, $A = \pi \dfrac{(0.6 \times 10^{-3})^2}{4}$

(remembering that 1 mm = $10^{-3}$ m).

$$R = \frac{430 \times 10^{-9} \times 6.5}{2.82 \times 10^{-7}} = 9.88 \text{ ohms, about 10 ohms.}$$

Using $R = 1.27 \times 10^{-3}\ \dfrac{L}{d^2}$

$$R = \frac{1.27 \times 10^{-3} \times 430 \times 6.5}{0.36} = 9.86 \text{ ohms, nearly 10 ohms.}$$

To find the length of wire needed for a given resistance value, the formula is transposed to

$$L = \frac{RA}{\rho},$$ using $R$ in ohms, $A$ in metres$^2$, and $\rho$ in ohm-metres

to obtain $L$ in metres. An alternative formula is

$$L = 785.4 \times \frac{Rd^2}{\rho}$$ with $R$ in ohms, $d$ in millimetres, $\rho$ in nano-

ohm metres.
To find the diameter of wire needed for a resistance $R$ and length $L$ metres, using $\rho$ in nano-ohm metres, the formula is

$$d = 3.57 \times 10^{-2} \sqrt{\frac{\rho L}{R}}$$ in millimetres

For some purposes, conductivity is used in place of resistivity. The conductivity, symbol $\sigma$ (Greek sigma), is defined as $\dfrac{1}{\text{resistivity}}$, so that $\rho = 1/\sigma$. The unit of conductivity is the siemens per metres, S/m. The resistivity formulae, using basic units, can be rearranged in terms of conductivity as follows

$$R = \frac{L}{\sigma A}, \qquad L = RA\sigma.$$

Conductivity values are also shown in *Table 1.4.*

*Resistor Construction*

The materials used for resistor construction are generally metal alloys, pure metal or metal oxide films, or carbon. Wirewound resistors use metal alloy wire wound onto ceramic formers. The winding must have a low inductance, so the wire is wound in the fashion shown in *Figure 1.3,* with each half of the winding wound in the opposite direction. Wire-wound resistors are used when very low values of resistance are needed, or when very precise values must be specified. Large resistance values, in the region of 20 k$\Omega$ to 100 k$\Omega$ need such fine-gauge wire that failure can occur due to corrosion in humid conditions, so that high-value wirewound resistors should not be used for marine or tropical applications unless the wire can be protected satisfactorily.

*Figure 1.3. Non-inductive winding of a wirewound resistor. The two halves of the total length of wire are wound in opposite directions so that their magnetic fields oppose each other.*

The majority of fixed resistors still use carbon composition, a mixture of graphite and clay whose resistivity value depends on the proportions of these materials used in the mixture. Because the resistivity value of such a mixture can be very high, greater resistance values can be obtained without the need for physically large components. Resistance value tolerances (see later) are high, however, because of the greater difficulty in controlling the resistivity of the mixture and the final dimensions of the carbon composition rod after heat treatment.

Metal film, carbon film and metal oxide film resistors are more recent types which are made by evaporating metals (in a vacuum) or tin oxide films (in air) onto ceramic rods. The resistance value is controlled (1) by controlling the thickness of the film; (2) by cutting a spiral pattern into the film after formation. These resistors are considerably cheaper to make than wirewound types, and can be made to closer tolerance values than carbon composition types.

Variable resistors and potentiometers can be made using all the methods that are employed for fixed resistors. The component is termed a potentiometer when connections are made to both ends as well as to the sliding connection; a variable resistor when only one end connection and the sliding connection is used. By convention, both are wired so that the quantity being controlled is *increased* by clockwise rotation of the shaft as viewed by the operator.

Any mass-production process aimed at producing one dimension will inevitably produce a range of values whose maximum tolerance can be specified. The tolerance is the maximum difference between any actual value and the target value, usually expressed as a percentage. For example a 10 k$\Omega$ 20% resistor may have a value of

$$10\,000 + \left(\frac{20}{100} \times 10\,000\right) = 12 \text{ k}\Omega \text{ or } 10\,000 - \left(\frac{20}{100} \times 10\,000\right) = 8 \text{ k}\Omega.$$

Tolerance series of preferred values (shown in *Table 1.5*), are ranges of target values chosen so that no component can be rejected. The mathematical basis of these preferred value figures is the sixth root of ten ($\sqrt[6]{10}$) for the E6 20% series (there are 6 steps of value between 1 and 6.8) and the twelfth root of 10 ($\sqrt[12]{10}$) for the E12 10% series. The figures produced by this series are rounded off. For example,

$$\sqrt[6]{10} = 1.46, \quad (\sqrt[6]{10})^2 = 2.15, \quad (\sqrt[6]{10})^3 = 3.16, \quad (\sqrt[6]{10})^4 = 4.64,$$
$$(\sqrt[6]{10})^5 = 6.8.$$

These figures are rounded to the familiar 1.5, 2.2, 3.3, 4.7, 6.8 used in the 20% series, and similar rounding is used for the 10% and 5% series.

A simpler view of the tolerance series is that, taking the 20% series as an example, 20% up on any value will overlap in value with 20% down on the next value. For example

4.7 + 20% = 5.64, and 6.8 − 20% = 5.44, allowing an overlap.

**Table 1.5. PREFERRED VALUES TOLERANCE SERIES**

| E6 series (20%) | E12 series (10%) | E24 series (5%) |
|---|---|---|
| 1.0 | 1.0 | 1.0 |
|  |  | 1.1 |
|  | 1.2 | 1.2 |
|  |  | 1.3 |
| 1.5 | 1.5 | 1.5 |
|  |  | 1.6 |
|  | 1.8 | 1.8 |
|  |  | 2.0 |
| 2.2 | 2.2 | 2.2 |
|  |  | 2.4 |
|  | 2.7 | 2.7 |
|  |  | 3.0 |
| 3.3 | 3.3 | 3.3 |
|  |  | 3.6 |
|  | 3.9 | 3.9 |
|  |  | 4.3 |
| 4.7 | 4.7 | 4.7 |
|  |  | 5.1 |
|  | 5.6 | 5.6 |
|  |  | 6.2 |
| 6.8 | 6.8 | 6.8 |
|  |  | 7.5 |
|  | 8.2 | 8.2 |
|  |  | 9.1 |

The numbers then repeat, but each multiplied by ten, up to 91 $\Omega$, then multiplied by 100 up to 910 $\Omega$ and so on.

After manufacture resistors are graded with the 1%, 5% and 10% tolerance values removed. The remaining resistors are sold as 20% tolerance. Because of this, it is pointless to sort through a bag of 20% 6K8 resistors hoping to find one which will be of exactly 6K8 — such values will have been removed in the first sorting process by the manufacturer. Electronic circuits are designed to make use of wide tolerance components as far as possible. When close tolerance components are specified, it will be for a good reason, and 20% tolerance components cannot be substituted for 10% or 5% types.

### Resistor characteristics

Important characteristics of resistor types include resistance range, usable temperature range, stability, noise level and temperature coefficient. Wirewound resistors are available in values ranging from fractions of an ohm (usually 0.22 $\Omega$) up to about 10 k$\Omega$; carbon compositions from about 2.2 $\Omega$ to 10 M$\Omega$, with film resistors available in ranges which are typically 1 $\Omega$ to 1 M$\Omega$. Typical usable temperature ranges are $-40°$ to $+105°$C for composition, $-55°$C to $+150°$C for metal oxide. Wirewound resistors can be obtained which will operate at higher temperatures (up to $300°$C) depending on construction and value of resistance. The stability of value means the maximum change of value which can occur during shelf life, on soldering, or in adverse conditions, particularly operation at high temperature in damp conditions. Composition resistors have poorest stability, with typical shelf life change of 5%, soldering change of 2% and 'damp heat' change of 15% in addition to normal tolerance. Metal oxide resistors can have shelf life changes of 0.1%, soldering change of 0.15%, 'damp heat' changes of 1%, typically. The noise level of a resistor is specified in terms of microvolts ($\mu$V) of noise signal generated for each volt of d.c. across the resistor, and range from 0.1 $\mu$V/V for metal oxide to a minimum of 2 $\mu$V/V for composition (increasing with resistance value for composition resistors). The formula generally used for noise level of carbon composition resistors is $2 + \log_{10} \dfrac{R}{1\,000}$ $\mu$V/V. For example, a 680 k$\Omega$ resistor would have a noise level of $2 + \log_{10} \dfrac{680\,000}{1\,000} = 2 + \log_{10} 680\,\mu$V/V $= 4.8\,\mu$V/V.

The temperature coefficient of resistance measures the change of resistance value as temperature changes. The basic formula is

$$R_\theta = R_0 \,(1 + a\,\theta)$$

where $R_\theta$ = resistance at temperature $\theta°$C, $R_0$ = resistance at $0°$C, $a$ = temperature coefficient, $\theta$ = temperature in $°$C.

The value of temperature coefficient is usually quoted in parts per million per $^\circ$C (ppm/$^\circ$C), and this has to be converted to a fraction (by dividing by one million) to use in the above formula.

*Example:* What is the value of a 6.8 k$\Omega$ resistor at 95$^\circ$C, if the temperature coefficient is +1 200 ppm/$^\circ$C?

*Solution:* Converting +1 200 ppm/$^\circ$C to standard form,

$$= \frac{+1\ 200}{1\ 000\ 000} = 1.2 \times 10^{-3} \qquad (0.0012)$$

Using the formula: $R_0 = 6.8\ (1 + 1.2 \times 10^{-3} \times 95)$

(The multiplication of $1.2 \times 10^{-3} \times 95$ must be carried out before adding 1).

$= 6.8\ (1 + 0.114)$

$= 7.57$ k$\Omega$

Note that if the resistance at some other temperature $\phi$ is given (as distinct from the resistance at 0$^\circ$C) then the formula becomes

$$R_0 = R\phi \left(\frac{1 + \alpha\theta}{1 + \alpha\phi}\right). \text{ The 1s cannot be cancelled.}$$

For example: if a resistor has a value of 10 $\Omega$ at 20$^\circ$C, its resistance at 80$^\circ$C can be found. If the value of $\alpha$ is $1.5 \times 10^{-3}$, then

$$R_0 = 10 \times \frac{1 + 80 \times 1.5 \times 10^{-3}}{1 + 20 \times 1.5 \times 10^{-3}} = 10 \times \frac{1 + .12}{1 + .03} = 10 \times \frac{1.12}{1.03} = 10.87\Omega$$

Temperature coefficients may be positive, meaning that the resistance will increase as the temperature rises, or negative meaning that the resistance will decrease as the temperature rises. Carbon composition resistors have temperature coefficients which vary from +1200 ppm/$^\circ$C, metal oxide types have the lowest temperature coefficient values of ±250 ppm/$^\circ$C.

## Dissipation and temperature rise

The dissipation rating, measured in watts (W), for a resistor indicates how much power can be converted to heat without damage to the resistor. The rating is closely linked to the physical size of the resistor, so that ¼ W resistors are much smaller than 1 W resistors of the same resistance value. These ratings assume 'normal' surrounding (ambient) temperatures, and for high temperature use, derating must be applied according to the manufacturer's specification. For example, a ½ W component may have to be used in place of a ¼ W component when the ambient temperature is 70$^\circ$C.

*Figure 1.4* shows the graph of temperature rise plotted against dissipated power for average ½ W and 1 W composition resistors. Note that these figures are of temperature rise *above* the ambient level. If such a temperature rise takes the resistor temperature above the maximum temperature permitted in its type, a higher wattage resistor must be used.

The dissipation in watts is given by $W = VI$ with $V$ the voltage across a conductor in volts and $I$ the current through the conductor in amps.

Figure 1.4. Temperature rise and power dissipation for typical carbon resistors. The temperature scale is in °C above surrounding (ambient) temperature. For example, in a room at 20°C, a ½W resistor dissipating 0.1 W will be at a temperature of 40°C

When current is measured in mA, then $VI$ gives power dissipation in milliwatts. This expression for dissipated power can be combined with Ohm's law when the resistance of $R$ of the conductor is constant, giving

$$W = V^2/R \quad \text{or} \quad I^2R$$

The result will be in watts for $V$ in volts and $R$ in ohms or $I$ in amps and $R$ in ohms. When $R$ is in k$\Omega$, $V^2/R$ gives W in milliwatts; when $I$ is in mA and $R$ in k$\Omega$, $W$ is also in milliwatts.

Note that *power* is energy transformed (from one form to another) per second. The unit of energy is the joule; the number of joules of energy dissipated is found by multiplying the power in watts by the time in seconds for which the power has been dissipated.

### Variables and laws

The *law* of a variable resistor or potentiometer must be specified in addition to the quantities specified for any fixed resistor. The potentiometer law describes the way in which resistance between the slider and one contact varies as the slider is rotated; the law is illustrated by plotting a graph of resistance against shaft rotation angle (*Figure 1.5*). A linear law potentiometer (*Figure 1.5a*) produces a straight line graph, hence the name linear. Logarithmic (log) law potentiometers are

*Figure 1.5. Potentiometer laws: (a) linear, (b) logarithmic. In the USA the word 'taper' is used in place of 'law', and 'audio' in place of 'log'. Broken lines show tolerance limits*

extensively used as volume controls and have the graph shape shown in *Figure 5.1b*. Less common laws are anti-log and B-law; specialised potentiometers with sine or cosine laws are available.

### Resistors in circuit

Resistors in circuit obey Ohm's law and Kirchhoff's laws. Ohm's law is written in its three forms as

$$V = RI; \qquad R = VI; \qquad I = V/R$$

where $V$ is voltage, $R$ is resistance, $I$ is current.

The units of these quantities are as shown in *Table 1.6*. These equations can be applied even to materials *which do not obey Ohm's law*, if the value of $R$ is known. Materials which do not obey Ohm's law do not have a *constant* value of resistance, but the relationships given above hold good. The usefulness of the equations is greatest when the resistance values of resistors are constant.

### Table 1.6.  OHM'S LAW AND UNITS

Forms of the law: $V = RI$; $R = V/I$; $I = V/R$

| Units of V | Units of R | Units of I |
|---|---|---|
| Volts, V | Ohms, $\Omega$ | Amps, A |
| Volts, V | Kilohms, k$\Omega$ | Milliamps, mA |
| Volts, V | Megohms, M$\Omega$ | Microamps, $\mu$A |
| Kilovolts, kV | Kilohms, k$\Omega$ | Amps, A |
| Kilovolts, kV | Megohms, M$\Omega$ | Milliamps, mA |
| Millivolts, mV | Ohms, $\Omega$ | Milliamps, mA |
| Millivolts, mV | Kilohms, k$\Omega$ | Microamps, $\mu$A |

Kirchhoff's laws relate to the conservation of voltage and current. In a circuit, the voltage across each series component can be added to find the total voltage. Similarly the total current entering a junction must equal the total current leaving the junction. These laws are illustrated in *Figure 1.6.*

*Figure 1.6. Kirchhoff's laws. The current law states that the total current leaving a circuit junction equals the total current into the junction — no current is 'lost'. The voltage law states that the driving voltage (or e.m.f.) in a circuit equals the sum of voltage drops (IR) around the circuit*

Current law :  Current in $R_3 = I_1 + I_2$

Voltage law :  $E_1 = R_1 I_1 + R_3(I_1 + I_2)$

$E_2 = R_2 I_2 + R_3(I_1 + I_2)$

$I$

$R_1$

$R_2$

$R_3$

$V$

$I$

$R_T$

$V$

$R_T = R_1 + R_2 + R_3$

(a)

$I$

$V$

$R_1$

$R_2$

$R_3$

$I$

$R_T$

$V$

$$\frac{1}{R_T} = \frac{1}{R_1} + \frac{1}{R_2} + \frac{1}{R_3}$$

For two resistors in parallel  $R_T = \dfrac{R_1 R_2}{R_1 + R_2}$

(b)

*Figure 1.7. Resistors in series (a) and in parallel (b)*

*Figure 1.7* shows the rules for finding the total resistance of resistors in series or in parallel. When a combination of series and parallel connections is used, the total resistance of each series or parallel group must be found first before finding the grand total.

The superposition theorem is of great value in finding the voltages and currents in a circuit with two or more sources of voltage. *Figure 1.8* shows an example of the theorem in use. One supply is selected and the circuit redrawn to show the other supply or supplies short circuited. The voltage and current caused by the first supply can then be calculated, using Ohm's law and the rules for combining parallel resistors. Each supply is treated in this way in turn, and finally the currents and voltages caused by each supply are added.

### SUPERPOSITION PRINCIPLE

*In any linear network, the voltage at any point is the sum of the voltages caused by each generator in the circuit. To find the voltage caused by a generator, replace all other generators in the circuit by their internal resistances, and use Ohm's law.* A linear network means an arrangement of resistors and generators, with the resistors obeying Ohm's law and the generators having a constant voltage output and constant internal resistances.

*Example*: In the network shown, find the voltage across the 2.2k$\Omega$ resistor.

In this network, there are two generators and three resistors. The generators might be batteries, oscillators, or other signal sources.

To find the voltage caused by the 6V generator, replace the 4V generator by its internal resistance of 0.5k$\Omega$. Using Ohm's law, and the potential divider equation, $V_1 = 1.736$ V

To find the voltage caused by the 4V generator, the 6V generator is replaced by its 1k$\Omega$ internal resistance. In this case, $V_2 = 2.315$ V

Now the total voltage in the original circuit across the 2.2k$\Omega$ resistor is simply the sum of these, 4.051 V.

*Figure 1.8. Using the superposition theorem. This is a simple method of finding the voltage across a resistor in a circuit where more than one source is present.*

*Figure 1.9. Using Thévenin's theorem. The potential divider (a) has an output voltage, with no load, of 4.04 V. It is equivalent to a 4.04 V source whose internal resistance is found by imagining the voltage supply short-circuited (b and c), so that the equivalent is as shown in (d). This makes it easy to find the output voltage when a current is being drawn. Similarly the bridge circuit (e) will have an open-circuit voltage, with R5 removed, of 1.17 V across X and Y (f), and the internal resistance between these points is found by imagining the supply short circuited (g). The combination of resistors in (g) is resolved (h) to give the single equivalent (i)*

Thévenin's theorem is, after Ohm's law, one of the most useful electrical circuit laws. The theorem states that any linear network (such as resistors and batteries) can be replaced by an equivalent circuit consisting of a voltage source with a resistance in series. The size of the equivalent voltage is found by taking the open circuit voltage between two points in the network, and the series resistance by calculating the resistance between the same two points in the network, assuming that the voltage source is short circuited. Examples of the use of Thévenin's theorem are shown in *Figure 1.9.*

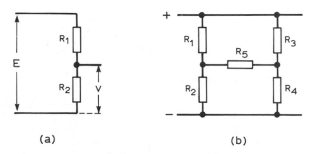

(a)                              (b)

*Figure 1.10. Potential divider (a) and bridge (b) circuits*

*Figure 1.10* shows two important networks, the potential divider and the bridge. When no current is taken from the potential divider, its output voltage is

$$V = \frac{R_2 E}{R_1 + R_2}$$

as shown, but when current is being drawn (as when a transistor is biased by this circuit) the equivalent circuit (using Thévenin's theorem) is more useful. The bridge circuit is said to be balanced when there is no voltage across $R_5$ (which is often a galvanometer or microammeter). In this condition,

$$\frac{R_1}{R_2} = \frac{R_3}{R_4}$$

If the bridge is *not* balanced, the equivalent circuit derived from Thévenin's theorem is, once again, more useful.

## Thermistors

Thermistors are resistors made from materials which have large values of temperature coefficients. Both p.t.c. and n.t.c. types are produced for applications ranging from measurement to transient current suppression. Miniature thermistors in glass tubes are used for temperature measurement, using a bridge circuit (*Figure 1.11*), for timing, or for stabilising the amplitude of sine wave oscillators (see Chapter 3). Such thermistors are self heating if the current through them is allowed to exceed the limits laid down by the manufacturers, so that the current flowing in a bridge measuring circuit must be carefully limited. Larger thermistor types, with lower values of cold resistance (at 20°C) are

*Figure 1.11. Thermistor bridge for temperature measurement*

*Figure 1.12. Graph of resistance plotted against temperature for typical thermistor*

used for current regulation, such as circuits for degaussing colour TV tubes or controlling the surge current into light bulbs or valve heaters. The form of graphs of resistance plotted against temperature is that of *Figure 1.12* and the formula for finding the resistance at any temperature is shown in *Figure 1.13* with examples.

$$R_{\theta 1} = R_{\theta 2} \cdot e^{B\left(\frac{1}{\theta_1} - \frac{1}{\theta_2}\right)}$$

$R_{\theta 1}$ – resistance at temperature $\theta_1$ (kelvins)

$R_{\theta 2}$ – resistance at temperature $\theta_2$ (kelvins)

$B$ – thermistor constant

Note: kelvin temperature $= °C + 273$

Calculator procedure:

Enter value of known temperature $\theta_1$ then $\boxed{\frac{1}{x}}$, $\boxed{-}$, enter value of $\theta_2$, $\boxed{\frac{1}{x}}$, $\boxed{=}$,

$\boxed{x}$, enter value of $B$ $\boxed{=}$ $\boxed{e^x}$ $\boxed{x}$ enter value of $R_{\theta 2}$

$\boxed{=}$ read answer

Example: A thermistor has a resistance of 47 kΩ at 20°C. what is its resistance at 100°C if its B value is 3900?

Equation is $R = 47 \times e^{3900\left(\frac{1}{373} - \frac{1}{293}\right)}$

$= 2 \cdot 7 kΩ$

*Figure 1.13. Finding thermistor resistance at any temperature, knowing the thermistor constant, B, and the resistance at a given temperature*

## Capacitors

Two conductors which are not connected and are separated by an insulator constitute a capacitor. When a cell is connected to such an arrangement, current flows *momentarily*, transferring charge (in the form of electrons) from one conducting plate (the + plate) to the other. When a quantity of charge $Q$ has been transferred, the voltage across the plates equals the voltage $V$ across the battery. For a given arrangement of conductors, the ratio $\frac{Q}{V}$ is a constant, and is called capacitance. The relationship can be written in three forms

$$Q = CV \qquad C = \frac{Q}{V} \qquad V = \frac{Q}{C}$$

The parallel-plate capacitor is the simplest practical arrangement, and its capacitance value is relatively easy to calculate. For a pair of parallel plates of equal area $A$, separation $d$, the capacitance is given by

$$C = \frac{\epsilon_r \epsilon_o A}{d}$$

The quantity $\epsilon_o$ is called the permittivity of free space and has the constant value of $8.84 \times 10^{-12}$ farads per metre.

Air has approximately this value of permittivity also, but other insulating materials have values of permittivity which are higher by the factor $\epsilon_r$, which is different from each material. Values of this quantity, the relative permittivity (formerly called dielectric constant) are shown in *Table 1.7*.

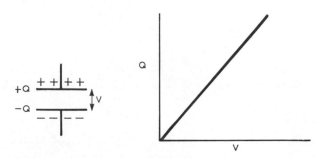

*Figure 1.14. Basic principles of the capacitor. The relationship Q/V, shown in the graph, is defined as the capacitance, C*

The formula can be recast, using units of $cm^2$ for area and mm for spacing, as

$$C = \frac{0.88 \times \epsilon_r \times A}{d} \text{ pF}$$

For small plate sizes, these units are more practical, but some allowance must be made for stray capacitance between any conductor and the metal surrounding it. Even a completely isolated conductor will have some capacitance.

Table 1.7.   VALUE OF $\epsilon_o$ = 8.84 pF PER METRE

| Material | Relative permittivity value |
| --- | --- |
| Aluminium oxide | 8.8 |
| Araldite resin | 3.7 |
| Bakelite | 4.6 |
| Barium titanate | 600—1200 (varies with voltage) |
| Magnesium silicate | 5.6 |
| Nylon | 3.1 |
| Polystyrene | 2.5 |
| Polythene | 2.3 |
| PTFE | 2.1 |
| Porcelain | 5.0 |
| Quartz | 3.8 |
| Soda glass | 6.5 |
| Titanium dioxide | 100 |

*Example 1:* Find the capacitance of two parallel plates 2 cm X 1.5 m, spaced by a 0.2 mm layer of material of relative permittivity 15.

*Solution:* Using $C = \dfrac{\epsilon_r \epsilon_o A}{d}$ $\epsilon_r = 15;\ \epsilon_o = 8.84 \times 10^{-12}$ F/m;

$A = 0.02 \times 1.5$ m$^2$; $d = 0.2 \times 10^{-3}$ m

$$C = \frac{15 \times 8.84 \times 10^{-12} \times 0.02 \times 1.5}{0.2 \times 10^{-3}} = 1.989 \times 10^{-8}\ \text{F}$$

about $2 \times 10^{-8}$ F or 0.02 $\mu$F

*Example 2:* Find the capacitance of two parallel plates 2 cm X 1 cm spaced 0.1 mm apart by a material with relative permittivity 8.

*Solution:* Using $C = \dfrac{0.88 \times \epsilon_r \times A}{d}$ with $A = 2 \times 1 = 2$ cm$^2$,

$d = 0.1$ mm

$$C = \frac{0.88 \times 8 \times 2}{0.1}\ \text{pF} = 140.8 \text{ or } 141 \text{ pF}$$

*Construction*

Small value capacitors can be made using thin plates of insulating material metallised on each side to form the conductors. Thin plates can be stacked and interconnected (*Figure 1.15c*), to form larger

(a)

$C = \dfrac{\epsilon_o A}{d}$
for air-spaced plates

(b)

Material with relative permittivity $\epsilon_r$

$C = \dfrac{\epsilon_r \epsilon_o A}{d}$

(c)

*Figure 1.15. The parallel-plate capacitor. The amount of capacitance is determined by the plate area, plate separation, and the relative permittivity of the material between the plates* capacitance values up to 1000 pF or more. Silver mica types are used when high stability of value is important as in oscillators; and ceramic for less important uses (such as decoupling). Ceramic tubular capacitors make use of small tubes silvered inside and outside.

Rolled capacitors use as dielectric strips of paper, polyethylene, polyester, polycarbonate or other flexible insulators which are metallised (by vacuum coating) on each side and then rolled up (*Figure 1.16*),

*Figure 1.16. The rolled construction used for capacitors which make use of sheet dielectrics such as paper, polyester, polystyrene or polycarbonate*

with another insulating strip to prevent the metallising on one side shorting against the metallising on the other side. Using this construction, quite large capacitance values can be achieved in small volume; up to a few $\mu$F is common.

Electrolytic capacitors are used when very large capacitance values are needed. One 'plate' is of aluminium in contact with an aluminium perborate solution in the form of a jelly or paste. The insulator is a film of aluminium oxide which forms on the positive 'plate' when a voltage, called the forming voltage, is applied during manufacture. Because the layer of oxide can be very thin (of only a few molecules thick) and the surface area of the aluminium can be made very large by roughening the surfaces, very large capacitance values can be achieved. The disadvantage of aluminium electrolytics include leakage current which is high compared to that of other capacitor types, polarisation (the + and − markings must be observed) and comparatively low voltage operation (less important when only transistor circuits are used). Incorrect polarisation can break down the oxide layer and if large currents can flow (as is the case in a power supply reservoir capacitor) the capacitor can explode, showering its surroundings with corrosive jelly.

Tantalum electrolytes can be used unpolarised (but not necessarily reverse polarised) and have much lower leakage currents that aluminium types, making them more suitable for some applications.

*Capacitor characteristics*

The same series of preferred values (20%) as is used for resistor values is also applied to values of capacitance (other than electrolytics) though older components will generally be marked with values such as 0.02 μF which are for all practical purposes equivalent to the preferred value 0.022 μF. Some manufacturers mark the values in pF only, using the rather confusing k to indicate thousands of pF (*Table 1.8*). Colour coded values are always in pF.

Electrolytic capacitors are always subject to very large tolerance values, of the order of −50% +100%, so that the actual capacitance value may range from half of the printed value to double that value. The insulation resistance between the plates is often so low that capacitance meters are unable to make accurate measurements. Capacitor values marked in circuit diagrams can use the BS 1852 method (6n8, 2μ2, etc) but are often marked in μF and pF. Quite commonly, fractions refer to μF and whole numbers to pF unless marked otherwise so that values of 0.02, 27, 1000, 0.05 mean 0.02 μF, 27 pF, 1000 pF and 0.05 μF respectively.

For all capacitors, the working voltage rating (abbreviated VW) must be carefully observed. Above this voltage, sparking between the

### Table 1.8. CAPACITOR COLOUR CODING

Most capacitors are marked with values in μF or pF. The letter k is sometimes used in place of nF, i.e. 10 k = 10 nF = 0.01 μF. Colour coding is sometimes used:

Bands A, B, C are used for coding values in pF in the same way as for resistors — remember that 1000 pF = 1 nF = 0.001 μF
Band D — Black = 20% White = 10%
Band E — Red — 250 V d.c. working
Yellow — 400 V d.c. working

Colour code for small block capacitors (mainly polyester)

Tantalum electolytic capacitors are also sometimes colour coded, but with values in μF rather than pF

| Band | 1 | 2 | 3 | 4 | |
|------|---|---|---|---|---|
| Black | — | 0 | ×1 | 10 V | |
| Brown | 1 | 1 | ×10 | | |
| Red | 2 | 2 | ×100 | |  |
| Orange | 3 | 3 | — | | |
| Yellow | 4 | 4 | — | 6.3 V | |
| Green | 5 | 5 | — | 16 V | |
| Blue | 6 | 6 | — | 20 V | |
| Violet | 7 | 7 | — | | |
| Grey | 8 | 8 | ×0.01 | 25 V | |
| White | 9 | 9 | ×0.1 | 3V | |
| Pink | | | | 35 V | |

conductors can break down the insulation, causing leakage current and eventual destruction of the capacitor. The maximum voltage that can be used is much lower at high temperatures than at low temperatures.

Values as low as 3 V may be found in high capacitance electrolytics; and values as high as 20 kV for ceramic capacitors intended for transmitters. The common voltage ranges used for capacitors in transistor circuits are shown in *Table 1.9.*

### Table 1.9. CAPACITORS – COMMON WORKING VOLTAGES

| 10 V | 16 V | 20 V | 25 V | 35 V | 40 V |
|------|------|------|------|------|------|
| 63 V | 100 V | 160 V | 250 V | 400 V | 1000 V |

Changes of temperature and of applied d.c. voltage affect the value of capacitors because of changes in the dielectric. Both p.t.c. and n.t.c. types can be obtained, and the two are often mixed to ensure minimal capacitance change in, for example, oscillator circuits. Paper and polyester capacitors have, typically, positive temperature coefficients of around 200 ppm/°C, but silver micas have much lower (positive) temperature coefficients. Aluminium electrolytics have large positive temperature coefficients, with a considerable increase in leakage current as temperature increases. In addition, electrolytics cannot be operated below about −20°C, as the electrolyte paste freezes. The normal working range for other types is −40°C to +125°C, though derating may be needed at the higher temperature. Voltage ratings are generally for 70°C working temperature.

A few types of capacitors, notably 'High-K' ceramics, change value as the applied voltage is changed. Such capacitors are unsuitable for use in tuning circuits and should be used only for non-critical decoupling and coupling applications.

Variable capacitors can make use of variation of overlapping area or of spacing between plates. Air dielectric is used for the larger types (360 pF or 500 pF), but miniature variables make use of mica or plastic sheets between the plates. Compression trimmers are manufactured mainly in the smaller values, up to 50 pF. In use, the moving plates are always earthed, if possible, to avoid changes of capacitance (due to stray capacitance) when the control shaft is touched.

### Energy and charge storage

The amount of charge stored by a capacitor is given by

$$Q = CV$$

When $C$ is in $\mu$F and $V$ in volts, $Q$ is in microcoulombs ($\mu$C).

*Example:*  How much charge is stored by a 0.1 $\mu$F capacitor charged
to 50 V?

*Solution:*  Using $Q = CV$ with $C$ in $\mu$F, $V$ in volts

$Q = 0.1 \times 50 = 5\ \mu$C.

When charged capacitors are connected to each other (but isolated from
a power supply), the total charge is constant, equal to the sum of all
charges on the capacitors. If the voltages are not equal, energy will be
lost (as electromagnetic radiation) when the capacitors are connected.
The amount of energy, in joules, stored by a charged capacitor is
most conveniently given by $W = \frac{1}{2}CV^2$.

Other equivalent expressions are $\frac{1}{2}\frac{Q^2}{C}$ or $\frac{1}{2}QV$.

*Example:*  How much energy is stored by a 5 $\mu$F capacitor charged
to 150 V?

*Solution:*  Using $\frac{1}{2}CV^2$, $C = 5 \times 10^{-6}$, $V = 150$

$W = \frac{1}{2} \times 5 \times 10^{-6} \times (150)^2$

$= 0.056$ J

This is used in connection with the use of capacitors to fire flash bulbs
or in capacitor discharge car ignition systems.

In circuits, the laws concerning the series and parallel connections of
capacitors are the *inverse* of those for resistors:

For capacitors in parallel: $C_{total} = C_1 + C_2 + C_3 + \ldots$

For capacitors in series: $\dfrac{1}{C_{total}} = \dfrac{1}{C_1}\ \dfrac{1}{C_2}\ \dfrac{1}{C_3}$

## Time Constants

The charging and discharging of a capacitor is never instant. When a
sudden step of voltage is applied to one plate of a capacitor, the other
plate will step by the same amount. If a resistor is present, connecting
the second plate to another voltage level, the capacitor will then charge
to this other voltage level. The time needed for this change is about
four time constants, as shown by *Figure 1.17*. The quantity, time
constant $T$, is measured by $R \times C$ where $R$ is the resistance of the
charge/discharge resistor and $C$ is the capacitance. For $C$ in farads and
$R$ in ohms, $T$ is in seconds. For the more practical units of $\mu$F and with
resistance in k$\Omega$, $T$ is in milliseconds (ms) or for $C$ in nF and resistance
in k$\Omega$, $T$ is in microseconds ($\mu$s).

*Example:*  In the circuit of *Figure 1.18*, how long does the voltage
at the output take to die away?

(a)

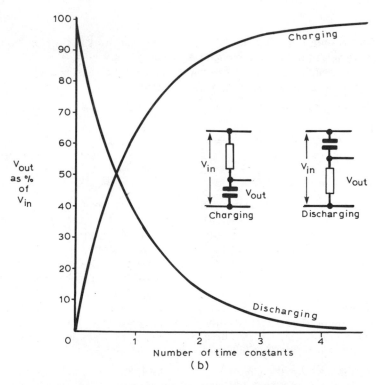

(b)

*Figure 1.17. Capacitor charging and discharging. (a) Principles of charging, (b) Universal charge/discharge curves*

*Figure 1.18. Time constant: differentiating circuit*

*Solution:*   With $C$ = 0.01 $\mu$F = 10 nF and $R$ = 15 k$\Omega$, $T$ = 150 $\mu$s.

Four time constants will be 4 × 150 $\mu$s = 600 $\mu$s, so that we can take it that the output voltage has reached zero after 600 $\mu$s.

*Example:*   In the circuit of *Figure 1.19*, how long does the capacitor take to charge to 10 V?

*Solution:*   With $C$ = 0.22 $\mu$F = 220 nF and $R$ = 6.8 k$\Omega$, $T$ = 6.8 × 220 = 1496 $\mu$s

Four time constants will be 4 × 1496 = 5984 $\mu$s or 5.98 ms, approx. 6 ms to charge.

*Figure 1.19. Time constant: integrating circuit*

Note that a time of four time constants is taken for charging or discharging, because in practice charging or discharging is virtually complete by this time. The shape of the graph, however, indicates that charge is still being moved even several hundred time constants later.

*Reactance*

The reactance of a capacitor for a sine wave signal is given by

$$X_c = \frac{1}{2 \pi f C} \qquad (2 \pi = 6.28)$$

where $C$ is capacitance in farads, $f$ is frequency in hertz.

Reactance is measured in ohms, and is the ratio

$$\frac{\widetilde{V}}{\widetilde{I}}$$

where $\widetilde{V}$ is a.c. voltage across the capacitor and $\widetilde{I}$ is the a.c. current in the circuit containing the capacitor.

Unlike resistance, reactance is not a constant but varies inversely with frequency (*Figure 1.20*). In addition, the current sine wave is ¼ cycle (90°) ahead of the voltage sine wave across the capacitor plates. For phase and amplitude graphs of *CR* circuits, see later this chapter.

(a)

(b)

*Figure 1.20. Capacitive reactance to a.c. signals. (a) Graph showing how capacitive reactance varies with frequency of signal. (b) Phase shift. As the capacitor charges and discharges, current flows alternately in each direction. The maximum current flow occurs when the capacitor is completely uncharged (zero voltage), and the maximum voltage occurs when the capacitor is completely charged (zero current). The graph of current is therefore ¼ cycle (90°) ahead of the graph of voltage*

## Inductors

An inductor is a component whose action depends on the magnetic field which exists around any conductor when a current flows through that conductor. When the strength of such a magnetic field (or magnetic flux) changes, a voltage is induced between the ends of the conductor. This voltage is termed an induced e.m.f., using the old term e.m.f. (electromotive force) to mean a voltage which has not been produced by current flowing through a resistor.

Faraday's Laws: Voltage induced depends on strength of magnet, speed of magnet (or coil) number of turns of coil, area of cross-section of coil

Magnet

Magnet

Coil

Coil

Magnet

Lenz's Law

Magnet moving away from coil- magnetic field attracts magnet

Magnet moving towards coil coil- magnetic field repels magnet

The direction of induced e.m.f is such that it always opposes the change ( movement in this case) which causes it

*Figure 1.21. Faraday's and Lenz's laws. Faraday's laws relate to the size of the induced (generated) voltage in a coil to the strength, speed of the magnet and the size of the coil. Lenz's law is used to predict the direction of the voltage*

If we confine out attention to static devices (coils and transformers rather than electric motors), then the change of magnetic field or flux can only be due to a change of current through one conductor. The induced e.m.f. is in such a direction that it opposes this change of current, and the faster the rate of change of current, the greater the opposing e.m.f. Because of its direction, the induced e.m.f. is called a *back e.m.f.* The laws governing these effects are Faraday's law and Lenz's law, summarised in *Figure 1.21.*

The size of the back e.m.f. can be calculated from the rate of change of current through the conductor and the details of construction of the conductor, straight wire or coil, number of turns of coil, use of a core etc. These constructional factors are lumped together as one quantity called inductance, symbol $L$. By definition,

$$E = L\frac{dI}{dt}$$

where $E$ is the back e.m.f., $L$ is inductance and $\frac{dI}{dt}$ is rate of change of current. The symbol 'd' in this context means 'change of' the quantity written after 'd'. If $E$ is measured in volts and $\frac{dI}{dt}$ in amperes per second, then $L$ is in henries (H).

*Example:*  What back e.m.f. is developed when a current of 3A is reduced to zero in $\frac{1}{50}$ s (20ms) through a 0.5 H coil?

*Solution:*  The amount of back e.m.f. is found from

$$E = L\frac{dI}{dt}$$

$$= 0.5 \times \frac{3}{20 \times 10^{-3}}$$

$$= 75 \text{ V}.$$

Note that this 75 V back e.m.f. will exist only for as long as the current is changing (20 ms), and may be *much* greater than the voltage drop across the coil when a steady current is flowing.

The rate of change of current is seldom uniform, so the back e.m.f. is usually a pulse waveform, whose maximum value must be found by measurement.

The existence of inductance in a circuit causes a reduction in the rate at which current can increase or decrease in the circuit. For a coil with inductance $L$ and resistance $R$ the time constant $T$ is $L/R$ seconds ($L$ in henries, $R$ in ohms). *Figure 1.22* shows how the current at a time $t$ after switch-on varies in an inductive circuit — once again we take the time of four time constants to represent the end of the process.

The large e.m.f. which is generated when current is suddenly switched off in an inductive circuit can have destructive effects, causing sparking at contacts or breakdown of transistor junctions. *Figure 1.23* shows the commonly used methods of protecting switch contacts and transistor junctions from these switching transients.

*Figure 1.22. The growth of current in an inductive circuit*

The changing magnetic field around one coil of wire will also affect windings nearby. The two windings are then said to have mutual inductance, symbol *M*. By definition

$$M = \frac{\text{back e.m.f. induced in second winding}}{\text{rate of change of current in first winding}}$$

using the same units as before, so that the unit of *M* is the henry.

*Figure 1.23. Protection against voltage surges in inductive circuits. (a) Using a capacitor across switch contacts, (b) using a diode across the inductor*

## Inductance calculations

Of all electronics calculations, those of inductance are the least precise. When an air-cored coil is used, the changing magnetic field does not affect all turns equally. Using a magnetic core makes the shape of the magnetic field more predictable, but makes its size less predictable. In addition, the permeability of the core changes considerably if d.c. flows in windings. Any equations for inductance are therefore very approximate, and should be used only as a starting point in the construction of an inductor. *Table 1.10* shows a formula for the number of

### Table 1.10.   INDUCTANCE OF A SINGLE LAYER SOLENOID

To find the number of turns needed for a given value of inductance, assuming air core

$$n = \frac{0.4Ld + 0.16L^2d^2 + 0.0033r^3L}{0.0031r^2}$$

n turns

where   $r$ = coil radius in mm
$L$ = inductance in μH
$d$ = wire diameter in mm
$n$ = number of close wound turns

turns of a single-layer close wound coil (solenoid) to achieve a given inductance. This approximate formula gives reasonable results for air-cored coils of values commonly used for radio circuit tuning. The addition of a core (generally of ferrite material) will cause an increase in inductance which could be by a factor as great as the relative permeability

### Table 1.11.   RELATIVE PERMEABILITY VALUES

Relative permeability, $\mu_r = \dfrac{\text{Inductance of coil with core}}{\text{Inductance of coil without core}}$

Alternatively, inductance value with core = $\mu_r$ × inductance value without core.

| Material | Relative permeability maximum value |
|---|---|
| Silicon-iron | 7 000 |
| Cobalt-iron | 10 000 |
| Permalloy 45 | 23 000 |
| Permalloy 65 | 600 000 |
| Mumetal | 100 000 |
| Supermalloy | 1 000 000 |
| Dustcores | 10 to 100 |
| Ferrites | 100 to 2 000 |

(*Table 1.11*) of the ferrite. The multiplying effect is seldom as large as the value of relative permeability, because the ferrite does not normally enclose the coil. Manufacturers of ferrite cores which enclose coils

*Example:*

Inductance of a 120 turn coil is measured as 840μH

How many turns need to be removed to give 500μH ?

Since: $L \propto n^2$ ( L - inductance, n - number of turns)

Then: $\dfrac{L_1}{L_2} = \dfrac{n_1^2}{n_2^2}$  and  $\dfrac{840}{500} = \dfrac{120^2}{n_2^2}$

$\therefore \quad n_2^2 = \dfrac{120^2 \times 500}{840} = 8571$

$\therefore \quad n = \sqrt{8571} = 93 \text{ approx}$

*Figure 1.24. Adjusting inductors to different values*

provide winding data appropriate for each type and size of core. *Figure 1.24* shows how inductors can be adjusted for a different inductance value, using the principle that inductance is proportional to the square of the number of turns.

### Inductive reactance

The reactance of an inductor for a sine wave signal of frequency *f* hertz is $2\pi fL$, where *L* is inductance in henries. The reactance is the ratio $\tilde{V}/\tilde{I}$

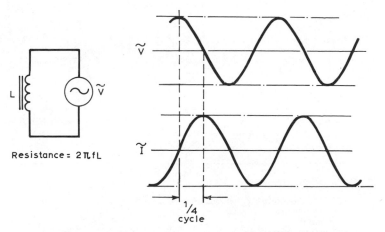

Resistance = $2\pi fL$

*Figure 1.25. Reactance and phase shift of a perfect (zero resistance) inductor*

($\widetilde{V}$ is signal voltage, $\widetilde{I}$ is signal current) and is measured in ohms. For a coil whose reactance is much greater than its resistance, the voltage sine wave is $90°$ (¼ cycle) ahead of the current sine wave (*Figure 1.25*).

## Untuned transformers

An untuned transformer consists of two windings, primary and secondary, neither or which is tuned by a capacitor, on a common core. For low frequency use, a massive core made from laminations (thin sections) of transformer steel alloy (such as silicon iron) must be used. Transformers which are used only for higher audio frequencies can make use of considerably smaller cores. At radio frequencies, the losses caused by transformer steels make such materials unacceptable, and ferrite materials are used as cores. For the highest frequencies, no core material is suitable and only self supporting air-cored coils (or pieces of straight wire or strip) can be used. In addition, high frequency currents flow mainly along the outer surfaces of conductors. This has two practical consequences — tubular conductors are as efficient as solid conductors (but use much less metal and can be water cooled) and silver plating can greatly decrease the effective resistance of a conductor.

For an untuned transformer with 100% coupling, the ratio of voltages $\widetilde{V}_s/\widetilde{V}_p$ is equal to the ratio of winding turns $N_s/N_p$, where $\widetilde{V}$ refers to a.c. voltage, $N$ to number of turns and $s$, $p$ to secondary, primary

### Table 1.12.    REACTIVE CIRCUIT RESPONSE

| Circuit | $\dfrac{V_{out}}{V_{in}}$ | Phase angle |
|---|---|---|
| | $\dfrac{1}{1 - f^2/f_o^2}$ | $0°$ when $f < f_o$ |
| | $\approx -f_o^2/f^2$ | $180°$ when $f > f_o$ |
| | $\dfrac{1}{1 - f_o^2/f^2}$ | $0°$ when $f > f_o$ |
| | $\approx -f^2/f_o^2$ | $180°$ when $f < f_o$ |

Notes:  $f_o$  is frequency of resonance $= \dfrac{0.16}{\sqrt{(LC)}}$

   $f$  is frequency at which response is to be found
   $>$  greater than
   $<$  less than
   $\approx$  approximately equal to

Circuits:

$$G = \frac{Vout}{Vin}$$

$\phi$ = phase angle

T = CR or L/R

$\omega = 2\pi \times$ frequency

| $\omega T$ | $\phi^\circ$ | G | $\omega T$ | $\phi^\circ$ | G |
|---|---|---|---|---|---|
| 0.1 | 84.3 | 0.099 | 0.9 | 48.0 | 0.67 |
| 0.2 | 78.7 | 0.196 | 1.0 | 45.0 | 0.707 |
| 0.3 | 73.3 | 0.287 | 1.5 | 33.7 | 0.83 |
| 0.4 | 68.2 | 0.37 | 2.0 | 26.6 | 0.9 |
| 0.5 | 63.4 | 0.45 | 3.0 | 18.4 | 0.95 |
| 0.6 | 59.0 | 0.51 | 4.0 | 14 | 0.97 |
| 0.7 | 55.0 | 0.57 | 5.0 | 11.3 | 0.98 |
| 0.8 | 51.34 | 0.62 | | | |

Circuits:

| $\omega T$ | $\phi^\circ$ | G | $\omega T$ | $\phi^\circ$ | G |
|---|---|---|---|---|---|
| 0.1 | −5.7 | 0.99 | 0.9 | −41.8 | 0.74 |
| 0.2 | −11.3 | 0.98 | 1.0 | −45 | 0.707 |
| 0.3 | −16.7 | 0.96 | 1.5 | −56 | 0.55 |
| 0.4 | −21.8 | 0.93 | 2.0 | −63 | 0.45 |
| 0.5 | −26.5 | 0.89 | 3.0 | −72 | 0.32 |
| 0.6 | −31 | 0.85 | 4.0 | −76 | 0.24 |
| 0.7 | −35 | 0.82 | 5.0 | −79 | 0.2 |
| 0.8 | −38.6 | 0.78 | | | |

*Figure 1.26. Amplitude/phase tables for LR and CR circuits. The quantity T is the time constant (CR or L/R), and $\omega$ is equal to $2\pi f$*

respectively. When an untuned transformer is used to transfer power between circuits of different impedance $Z_p$, $Z_s$, then the best match (maximum power transfer) condition is

$$\sqrt{\frac{N_s}{N_p}} = \frac{Z_s}{Z_p}$$

## LCR circuits

The action of CR and LR circuits upon a sine wave signal is to change both the amplitude and the phase of the signal. Universal amplitude/ phase tables can be prepared, using the time constant T of the CR and

*LR* circuit and the frequency *f* of the sine wave. These tables are shown, with examples, in *Figure 1.26*.

When a reactance (*L* or *C*) is in circuit with a resistance *R*, the general formula for the total impedance (*Z*) are as shown in *Table 1.12*. Impedance is defined as $Z = \tilde{V}/\tilde{I}$, but the phase angle between *V* and *I* will not be 90°, and will be 0° only when resonance (see later) exists.

The combination of inductance and capacitance produces a tuned circuit which may be series (*Figure 1.27a*) or parallel (*Figure 1.27b*). Each type of tuned (or resonant) circuit has a frequency of resonance, symbol, $f_o$, at which the circuit behaves like a resistance so that there is no phase shift between voltage and current. At other frequencies, the circuit may behave like an inductor or like a capacitor. Below the frequency of resonance, the parallel circuit behaves like an inductor, the series circuit behaves like a capacitor. Above the frequency of resonance the parallel circuit behaves like a capacitor, the series circuit like an inductor. At resonance, the parallel circuit behaves like a large value resistor and the series circuit like a small value resistor.

The series resonant circuit can provide voltage amplification at the resonant frequency when the circuit of *Figure 1.28* is used. The amount of voltage amplification is given by $\dfrac{2\pi fL}{R}$ or $\dfrac{1}{2\pi fCR}$ at the frequency of

Figure 1.27. Tuned circuits (a) series, (b) parallel

(a)          (b)

Figure 1.28. Voltage amplification of a tuned series circuit. The amplification is of the resonant frequency only, and can occur only if the signal source is of comparatively low impedance

resonance, and this quantity is often termed the circuit magnification factor, $Q$. There is no *power* amplification, as the voltage step up is achieved by increasing the current through the circuit, assuming that the signal input voltage is constant. *Table 1.12* shows typical phase and amplitude response formulae in universal form for the series resonant circuit. Note that the tuning capacitance may be a stray capacitance.

The parallel resonant circuit is used as a load which is a pure resistor (with no phase shift) only at the resonant frequency, $f_o$. The size of the equivalent resistance is called dynamic resistance and is calculated by the formula shown in *Figure 1.29*. The effect of adding resistors in

At resonance: $R_d = \dfrac{L}{CR}$

L = inductance in henries
C = capacitance in farads
R = resistance of coil in ohms

*Figure 1.29. Dynamic resistance of a parallel resonant circuit*

parallel with such a tuned circuit is shown in *Figure 1.30*, the dynamic resistance at resonance is reduced, but the resistance remains fairly high over a greater range of frequency. Such 'damping' is used to extend the bandwidth of tuned amplifiers. *Table 1.12* shows the amplitude and phase response of a parallel tuned circuit in general form. Once again, the tuning capacitance may be a stray capacitance.

*Figure 1.30. The effect of damping resistance on the resonance curve*

The impedance of a series circuit is given by the formula shown, with example, in *Table 1.12a*. Note that both amplitude (in ohms) and phase angle are given. The corresponding expression for a parallel circuit in which the only resistance is that of the coil $(R)$ is also shown in *Table 1.13b*. When a damped parallel circuit is used, the resistance of the coil has generally a negligible effect compared to the damping resistor, and the formula of *Table 1.13c* applies.

## Coupled tuned circuits

When two tuned circuits are placed so that their coils have some mutual inductance $M$, the circuits are said to be coupled. The size of the

## Table 1.13. IMPEDANCE Z AND PHASE ANGLE $\phi$

| Circuit | Z | $\phi$ | |
|---|---|---|---|
| | $Z = \sqrt{\left[ R^2 + \left( \omega L - \dfrac{1}{\omega C} \right)^2 \right]}$ | $\phi = \tan^{-1} \dfrac{\left( \omega L - \dfrac{1}{\omega C} \right)}{R}$ | (a) |
| | $Z = \sqrt{\left[ \dfrac{R^2 + \omega^2 L^2}{(1 - \omega^2 LC) + \omega^2 C^2 R^2} \right]}$ | $\phi = \tan^{-1} \dfrac{\omega \left[ L(1 - \omega^2 LC) - CR^2 \right]}{R}$ | (b) |
| | $Z = \dfrac{1}{\sqrt{\left[ \left( \dfrac{1}{R} \right)^2 + \left( \omega C - \dfrac{1}{\omega L} \right)^2 \right]}}$ | $\phi = \tan^{-1} R \left( \dfrac{1}{\omega L} - \omega C \right)$ | (c) |

**Notes:** $\omega = 2\pi \times$ frequency

$\tan^{-1}$ = angle whose tangent is equal to

mutual inductance is not simple to calculate; one approximate method is shown in *Table 1.14*. When the mutual inductance ($M$) between the coils is small compared to their self inductances ($L_1$, $L_2$) then the coupling is said to be loose, and the response curve shows a sharp peak. When the mutual inductance between the coils is large compared to their self inductances, the coupling is tight (or overcoupled) and the response curve shows twin peaks. For each set of coupled coils there is an optimum coupling at which the peak of the response curve is

### Table 1.14. MUTUAL INDUCTANCE

1. From values of coil size $S$ and $X$ (note that these are lengths divided by coil diameter) find $K$.
2. Knowing inductance values, $L_1, L_2$, find $M$.
   $M = k\sqrt{(L_1 L_2)}$

$S$ = spacing/diamater
$X$ = coil winding length/diameter

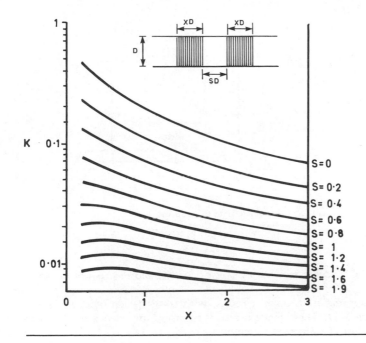

flattened and the sides steep. This type of response is an excellent compromise between selectivity and sensitivity.

The coefficient of coupling $k$ is defined by

$$k = \frac{M}{\sqrt{(L_1 L_2)}}$$

(or $\frac{M}{L}$ if both coils have the same value of $L$) and critical coupling occurs when $k = \frac{1}{Q}$, assuming that both coils have the same $Q$ factor — if they do not, then $Q = \sqrt{(Q_1 Q_2)}$. The size of the coefficient of coupling depends almost entirely on the spacing between the coils and no formulae are available to calculate this quantity directly.

Circuits

Approximate formulae

$$k = \frac{-\sqrt{(C_1\, C_3)}}{C_2}$$

(a)

$$k = \frac{-C_2}{\sqrt{(C_1\, C_3)}}$$

(b)

$$k = \frac{-\sqrt{(C_1\, C_3)}}{C_2}$$

k is coefficient of coupling
defined as $\dfrac{M}{\sqrt{(L_1 L_2)}}$

(c)

*Figure 1.31. Other methods of circuit coupling, and their design formulae*

Other types of coupled circuits, with some design data, are shown in *Figure 1.31*. These make use of a common impedance or reactance for coupling and are not so commonly used.

## Quartz crystals

Quartz crystals, cut into thin plates and with electrodes plated onto opposite flat faces, can be used as resonant circuits, with $Q$ values ranging from 20 000 to 1 000 000 or more. The equivalent circuit of a crystal is shown in *Figure 1.32*. The crystal by itself acts as a series

L, $C_C$, R equivalent values of crystal

$C_S$ stray capacitance across terminals

*Figure 1.32. Equivalent circuit of a quartz crystal*

resonant circuit with very large inductance, small capacitance and fairly small resistance (a few thousand ohms). The stray capacitance across the crystal will also permit parallel resonance at a frequency slightly higher than that of the series resonance. *Figure 1.33* shows how the

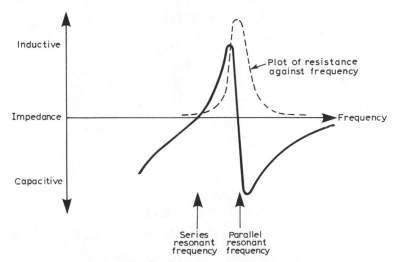

*Figure 1.33. Variation of reactance and resistance of a crystal near its resonant frequencies*

reactance and the resistance of a crystal vary as frequency is varied — the reactance is zero at each resonant frequency, the resistance is maximum at the parallel resonance frequency. Usually the parallel or the series frequency is specified when the crystal is manufactured.

### Wave filters

Wave filter circuits are networks containing reactive components ($L, C$) which accept or reject frequencies above or below cut-off frequencies (which are calculated from the values of the filter components). Output amplitude and phase vary considerably as the signal frequency approaches a cut-off frequency, and the calculations involved are beyond the scope of this book. Much more predictable response can be obtained, for audio frequencies at least, by using active filters (see Chapter 3).

### Measuring R, L, C

Resistance measurements can be made using the multimeter; the scale is non-linear but readings can be precise enough to indicate whether or not the value is within tolerance. Measurements of resistance, capacitance, self and mutual inductance can be carried out using bridge circuits (*Figure 1.34*) which rely on a 'null reading'. This means that potentiometers or switches are adjusted until a meter reading reaches a minimum,

At balance:
$$\frac{R_1}{R_2} = \frac{R_3}{R_4}$$

(a)

At balance:
$$C_2 = C_1 \frac{R_1}{R_2}$$

$$R_4 = R_3 \frac{R_2}{R_1}$$

(b)

At balance:
$$L_1 = \frac{R_1 R_4 C_1}{1 + \omega^2 C_1^2 R_3^2}$$

$$Q = \frac{1}{\omega C_1 R_3}$$

$\omega = 2\pi \times$ frequency of supply

(c)

*Figure 1.34. Measuring bridge circuits, with balance conditions. (a) Simple Wheatstone resistance bridge, (b) capacitance bridge, (c) inductance bridge* upon which the value of the quantity being measured can be read from the dials. Such bridge circuits use an audio frequency oscillator as a source of bridge voltage.

Direct-reading capacitance meters use a rather difference principle. Referring to *Figure 1.35*, the capacitor C is charged to a known voltage V, and then discharged through the meter, M. The amount of charge

*Figure 1.35. Principle of a direct-reading capacitance meter*

passing through the meter on each discharge is CV, so that if the switch is actuated f times per second, the amount of charge flowing per second is fCV. This is the average current $i$, read by the meter, so that $I = fCV$ or $C = \dfrac{I}{fV}$. By a suitable choice of switching frequency, charging voltage and meter range, values of capacitance ranging from 10 pF to several $\mu$F can be measured, though erratic results are sometimes experienced with electrolytic capacitors. The switching is carried out by transistors, as the switching speeds needed are beyond the range of mechanical switches, even reed switches.

# Chapter 2

# Active Discrete Components

## Diodes

Semiconductor diodes may use two basic types of construction, point contact or junction. Point contact diodes are used for small signal purposes where a low value of capacitance between the terminals is important — their main use nowadays is confined to r.f. demodulation. Junction diodes are obtainable with much greater ranges of voltage and current and are used for most other purposes. Apart from diodes intended for specialised purposes, such as light-emitting diodes, the fabrication materials are silicon or germanium, with germanium used almost exclusively for point contact demodulator diodes.

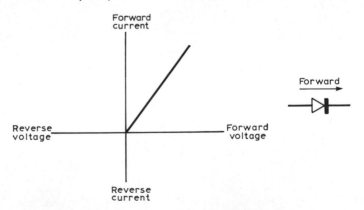

*Figure 2.1. Characteristic of an ideal diode.*

An ideal diode would conduct in one direction only, and the characteristic (the graph of current plotted against applied voltage) would look as in *Figure 2.1*. Practical diodes have a low forward resistance (not a *constant* value, however) and a high reverse resistance; and they conduct when the anode voltage is a few hundred millivolts higher than the cathode voltage.

The diode can be destroyed by excessive forward current, which causes high power dissipation at the junction or contact, or by using excessive reverse voltage, causing junction breakdown (see later), allowing it to conduct. Because reverse voltages are much higher than the voltage across a forward conducting diode, breakdown causes excessive current to flow so that once again the junction or contact is destroyed by excessive dissipation. For any diode therefore, the published ratings of peak forward current and peak reverse voltage should not be exceeded at any time, and should not be approached if reliable operation is to be achieved.

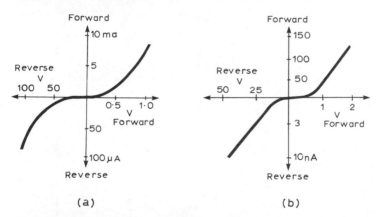

(a)  (b)

*Figure 2.2. Characteristics of real diodes. (a) Germanium point diode, (b) silicon junction diode. Note the different scales which have to be used to allow the graphs to be fitted into a reasonable space*

Characteristics for a typical point contact germanium diode and a typical small signal silicon junction diode are also shown in *Figure 2.2.* Comparing the two type of diode:

(a)  Germanium point contact diodes have lower reverse resistance values, conduct at a lower forward voltage (about 0.2 V) but have higher forward resistance because of their small junction area. They also have rather low peak values of forward current and reverse voltage.

(b)  Silicon junction diodes have very high values of reverse resistance, conduct at around 0.55 V forward voltage, can have fairly low forward resistance, and have fairly high peak values of forward current and reverse voltage.

The foward resistance of a diode is not a fixed quantity but is (very approximately) inversely proportional to current. Another approximation, useful for small currents, is that the forward voltage of a silicon diode increases by only 60 mV for a tenfold increase in current.

The effect of temperature change on a silicon diode is to change the forward voltage at any fixed value of current. A change of about 2.5 mV per $^\circ$C is a typical figure with the voltage reducing as the temperature is raised. The reverse (leakage) current is much more temperature dependent, and a useful rule of thumb is that leakage current doubles for each $10^\circ$C rise in temperature.

Zener diodes are used with reverse bias, making use of the breakdown which occurs across a junction when the reverse voltage causes a large electrostatic field across the junction. This breakdown limit occurs at low voltages (below 6 V) when the silicon is very strongly doped, and this type of breakdown is called zener breakdown. For such

*Figure 2.3. Zener diode. The true zener effect causes a 'soft' breakdown (a) at low voltages, the avalanche effect causes a sharper turnover (b)*

a true zener diode, the reverse characteristic is as shown in *Figure 2.3a*. As the graph shows, the reverse current does not suddenly increase at the zener voltage, and the voltage across the diode is not truly stabilised unless the current is more than a few milliamps. This type of characteristic is termed a 'soft' characteristic. In addition to this, a true zener diode has a negative temperature coefficient — the voltage across the diode (at a constant current value) decreases as the junction temperature is increased.

Avalanche breakdown occurs in diodes with lower doping levels, at voltages above about 6 V. The name is derived from the avalanche action in which electrons are separated from holes by the electric field across the junction, and these electrons and holes then cause further electron-hole separation by collision. These diodes have 'hard' characteristics (*Figure 2.3b*) with very little current flowing below the avalanche voltage and large currents above the avalanche voltage. In addition, the temperature coefficient of voltage is positive, so that the voltage across the diode *increases* as the junction temperature is raised.

Both types of diodes are known, however, as zener diodes, and these with breakdown voltages between 4 V and 6 V combine both effects. At a breakdown voltage of around 5.6 V, the opposing temperature characteristics balance, so that the breakdown voltage of a 5.6 V diode is practically unaffected by temperature. The stabilisation of the diode is measured by its dynamic resistance, defined as the ratio

$$\frac{\text{Voltage change, } V}{\text{Current change, } I} \text{ , units of ohms,}$$

when $V$ is the change of voltage across the diode caused by a change of current $I$ when the diode is stabilising.

This ratio should be below 50 ohms, and reaches a minimum value of about 4 ohms for a diode with a breakdown voltage of about 8 V.

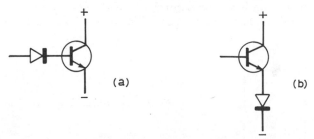

Figure 2.4. *Protecting the base-emitter junction of a transistor against excessive reverse voltage, (a) diode in base circuit, (b) diode in emitter circuit. The base circuit is preferred, since the diode does not have to pass the emitter current of the transistor*

Note that the base-emitter junction of many types of silicon transistors will break down by avalanche action at voltages ranging from 7 V to 20 V reverse bias, though this action does not necessarily cause collector current to flow. The base-emitter junction can be protected by a silicon diode (with a high breakdown voltage) wired in series (*Figure 2.4*).

Reference diodes are doped to an extent which makes the breakdown voltage practically constant despite changes in temperature. Voltages of 5 V to 6 V are used, and temperature changes ranging from ± 0.01% per degree to 0.000 5% per degree can be achieved.  Reference diodes are used for very precise voltage stabilisation.

## Varactor Diodes

All junction diodes have a measurable capacitance between anode and cathode when reverse biased, and this capacitance varies with voltage, being least when high reverse voltage are used. This variation is made use of for varactor diodes, in which the doping is arranged for the maximum possible capacitance variation. A typical variation is of 10 pF at 10 V bias to 35 pF at 1 V reverse bias. Varactor diodes are used for electronic tuning applications; a typical circuit is shown in *Figure 3.26* (Chapter 3).

## LEDs

Light emitting diodes use compound semiconductors such as gallium arsenide or indium phosphide. When forward current passes, light is emitted from the junction. The colour of the light depends on the material used for the junction, and the brightness is approximately proportional to forward current. LEDs have higher forward voltages when conducting; 1.6 to 2.2 V as compared to the 0.5 to 0.8 V of a silicon diode. The maximum permitted reverse voltages are low, typically 3 V, so that a silicon diode must be connected across the LED as

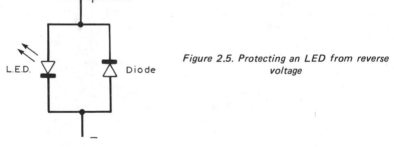

Figure 2.5. Protecting an LED from reverse voltage

shown in *Figure 2.5* if there is any likelihood of reverse voltage (or a.c.) being applied to the diode. A series resistor must always be used to limit the forward current unless pulsed operation is used.

For more specialised applications, microwave diodes of various types can be obtained which emit microwave radiation when forward biased and enclosed in a suitable resonant cavity. Tunnel diodes are diodes with an unstable portion of the characteristic (a reverse slope, indicating 'negative resistance'). When the tunnel diode is biased to the unstable region, oscillations are generated at whatever frequency is permitted by the components connected to the diode (*RC, LC,* cavity, etc.)

## Diode circuits

*Figure 2.6* shows some application circuits for diodes, with approximate design data where appropriate. Diode types should be selected with reference to the manufacturers data sheets, having decided on the basic reverse voltage and load current quantities required by the circuit.

*Figure 2.6. Some diode applications: (a) amplitude demodulation, (b) ratio detector for f.m. (c) signal clipping, (d) d.c. restoration, (e) bridge rectification*

## Transistors

Like signal diodes, transistors may be constructed using either silicon or germanium, but virtually all transistors now use silicon. The design data of this section refer to silicon transistors only.

The working principle of a transistor is that current flows between the collector and emitter terminals only when current is flowing between the base and emitter terminals. The ratio of these currents is called

(a)

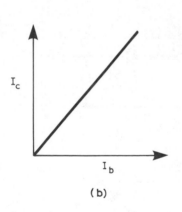

(b)

*Figure 2.7. Forward current transfer ratio (a) measuring circuit, (b) graph. The slope of the graph ($I_c/I_b$) is equal to the forward current transfer ratio, $h_{fe}$*

forward current transfer ratio, symbol $h_{fe}$. For the arrangement of *Figure 2.7*, the ratio is

$$h_{fe} = \frac{i_c}{i_b}$$

In databooks, a distinction is made between $h_{FE}$, for which $I_c$ and $I_b$ are d.c. quantities; and $h_{fe}$, for which $i_c$ and $i_b$ are a.c. quantities. The two quantities are however generally close enough in value to be interchangeable, and the symbol $h_{fe}$ will be used here to indicate both values. The size of $h_{fe}$ for any transistor can be measured in the circuit of *Figure 2.7a*; a simpler method, used in many transistor testers, is shown in *Figure 2.8*. Values vary from about 25 (power transistors operating at high currents) to over 1000 (some high-frequency amplifier types).

Base current will not flow unless the voltage between the base and the emitter is correct. The precise voltage at which current starts to

(a)

(b)

*Figure 2.8. A simple transistor tester (a) and its calibration graph (b)*

become detectable varies from one specimen (even of the same type number) of transistor to another, but for silicon transistors is generally about 0.5 V. The pnp type of transistor will require the emitter to be at a more positive voltage than the base, the npn type will require the base to be more positive than the emitter. When the transistor has the correct d.c. currents flowing (with no signal applied) it is said to be correctly biased. Amplification is carried out by adding a signal voltage to the steady voltage at the input of the transistor. The vast majority of transistor circuits use the base as the input terminal, though a few (common base amplifiers) use the emitter as an input terminal.

*Bias for linear amplifiers*

A linear amplifier produces at its output a waveform which is a perfect copy, but of greater amplitude, of the waveform at the base. The voltage gain of such an amplifier is defined as

$$G = \frac{V_{out}}{V_{in}}$$

where $V$ indicates a.c. (signal) voltage measurements. If the output waveform is not a perfect copy of the input, then the amplifier exhibits distortion of one sort or another. One type of distortion is non-linear distortion, in which the shape of the waveform is changed by the 'copying' process. Such non-linear distortion is caused by the action of the transistor and can be minimised by careful choice of transistor type (see later) and by correct bias.

A transistor is correctly biased when the desired amount of gain can be obtained with minimum distortion. This is easiest to achieve when the (peak to peak) output signal from the amplifier is much smaller than the supply voltage. This may be achieved with the no-signal voltage at the collector of the transistor at almost any reasonable level, but to allow for unexpected signal overloads, the preferred collector voltage is half-way between supply positive voltage and the voltage of the emitter.

When the value of collector resistor has been chosen, bias is applied by passing current into the base so that the collector voltage drops to the desired values of 0.5 $V_{ss}$ where $V_{ss}$ is supply voltage. For any bias system, the desired base current must be equal to

$$\frac{0.5\ V_{ss}}{R_L \times h_{fe}}$$ mA, with $V_{ss}$ in volts, $R_L$ in k$\Omega$, $h_{fe}$ as a ratio.

*Figure 2.9* shows three bias systems, with design data for obtaining a suitable bias voltage. The method of *Figure 2.9a* is the most difficult to use, as a different resistor value must be chosen for each transistor used. It may be necessary to use resistors in series or in parallel to achieve the correct value and the collector voltage will decrease noticeably as the temperature of the transistor increases. The method of *Figure 2.9b* is a considerable improvement over that of *Figure 2.9a*. The bias system may be designed around an 'average' transistor (with an average value of $h_{fe}$ for that type) and can then be used unchanged for other transistors without too serious a change in the collector voltage. In addition the collector voltage changes much less as the temperature changes.

The bias system of *Figure 2.9c* is one which can be used for any transistor provided that the current flowing through the two base bias resistors $R_1$, $R_2$ is much greater than the base current drawn by the transistor. Unlike the other two systems, the design formula does not

$$V_c = V_s - \frac{R_L h_{fe}(V_s - 0.6)}{R_1}$$

(a)

$$V_c = \frac{V_s - \dfrac{0.6 R_L h_{fe}}{R_1}}{1 + \dfrac{R_L h_{fe}}{R_1}}$$

(b)

$$V_b = \frac{V_s \times R_2}{R_1 + R_2}$$

$$V_e = V_b - 0.6$$

$$I_c = V_e / R_e$$

$$V_s = V - R_L I_c$$

(c)

*Figure 2.9. Transistor bias circuits: (a) simple system, usually unsatisfactory; (b) using negative feedback of bias and signal; (c) potential divider method*

require the $h_{fe}$ value of the transistor to be known if the standing current through the transistor is to be only a few milliamps. For power transistors, the quantities that are needed are the $V_{be}$ and $I_{be}$ at the bias current required. This system does not, however, stabilise the collector voltage so effectively against changes caused by changes of temperature.

### Transistor parameters and linear amplifier gain

Transistor parameters are measurements which describe the action of the transistor. The name parameter is used to distinguish these quantities from constants. Transistor parameters are not generally constants, they vary from one transistor to another (even of the same type) and from one value of bias current to another. One such parameter, the common-emitter current gain, $h_{fe}$, has already been described.

Of the parameters for linear amplifiers, $G_m$ is probably the most useful. $G_m$, called mutual conductance and measured in units of milli-siemens (mS) (equal to milliamps per volt) is defined by

$$G_m = \frac{\tilde{I}_c}{\tilde{V}_{be}}$$  where $I_c$ = a.c. signal current, collector to emitter

where $V_{be}$ = a.c. signal voltage, base to emitter.

The usefulness of $G_m$ as a parameter arises from the fact that the voltage gain of a transistor amplifier for small signals is given by

$$G = G_m R_L$$  where $R_L$ is the load resistance for signals (if $G_m$ is measured in mS and $R_L$ in k$\Omega$, gain is correctly specified)

Note that $R_L$ will generally be less than the resistance connected between the collector and the positive supply, because this value will be shunted by any other resistors connected through a capacitor to the collector

Current generator has infinite resistance and generates a signal current equal to $G_m \tilde{V}_{be}$

Figure 2.10. A useful equivalent circuit for the transistor. The signal voltage $\tilde{V}_{be}$ between the base and the emitter causes an output signal current $G_m \tilde{V}_{be}$. This current flows through the parallel resistor $R_C$ (the transistor output resistance), $R_L$, the load resistor, and any other load resistors in the circuit

(*Figure 2.10*). The input resistance of the next transistor (if used) will also be in parallel with the collector resistor.

A graph of collector current plotted against base-emitter voltage is not a straight line, so that the ratio $\frac{i_c}{V_{be}}$ which is $G_m$, is not constant. A useful rule of thumb for small bias currents is that $G_m$ = 40 X bias current in milliamps with $G_m$ in mS (mA/V). The shape of the graph is always curved for low currents, but can vary in shape at higher currents. For a few transistors, the $i_c$, $V_{be}$ graph has a noticeably straight portion, making these transistors particularly suitable for linear amplification applications. It is this straightness of the $G_m$ characteristic which makes some types of power transistor much more desirable (and high priced) for audio output stages than others.

To take advantage of these linear characteristics, of course, the bias must be arranged so that the working point is at the centre of the linear region with no signal input. 'Working point' in this context means the combination of collector volts and base volts which represents a point on the characteristic.

Two other useful parameters for silicon transistors used in the common-emitter circuit are the input and output resistance values. The input resistance is defined as

$$\frac{\text{signal voltage, base input}}{\text{signal current into base}}$$

with no signal at the collector, symbol $h_{ie}$.

Then output resistance is, using a similar definition

$$\frac{\text{signal voltage at collector}}{\text{signal current at collector}}$$

with no signal at the base, symbol $h_{oe}$.

The output resistance $h_{oe}$ has about the same range of values, from 10 k$\Omega$ to 50 k$\Omega$ for a surprisingly large number of transistors, irrespective of operating conditions, provided these are on the flat part of the $V_{ce}/I_{ce}$ characteristic (*Figure 2.11*). An average value of 30 k$\Omega$ can usually be taken.

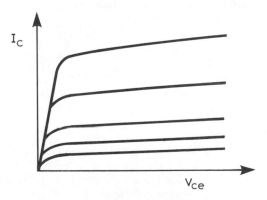

*Figure 2.11. The $I_c/V_{ce}$ characteristic. The flat portion is the operating part. The small amount of slope indicates that the output resistance, $R_c$, is high, usually 40 k$\Omega$ or more*

The input resistance is not a constant, because the input of a transistor is a diode, the base-emitter junction. The value of input resistance $h_{ie}$ is related to steady bias current, and to the other parameters

$$h_{ie} = \frac{h_{fe}}{G_m} \quad \text{and since } G_m = 40I_c,$$

$$= \frac{h_{fe}}{40I_c} \quad \text{where } I_c \text{ is the steady, no-signal, bias collector current.}$$

For example: if a transistor has an $h_{fe}$ value of 120, and is used at a current of 1 mA, then the input resistance (in k$\Omega$) is

$$h_{ie} = \frac{h_{fe}}{40I_c} = \frac{120}{40 \times 1} = 3\text{k}\Omega.$$

## Noise

Any working transistor generates electrical noise, and the greater the current flowing through the transistor the greater the noise. For bipolar transistors, the optimum collector current for low noise operation is given approximately (in milliamps) by

$$I_c = \frac{28\sqrt{h_{fe}}}{R_g} \quad \text{where } R_g \text{ is the signal source resistance in ohms.}$$

Low-noise operation is most important for the first stages of audio preamplifiers and for r.f. tuner and early i.f. stages. The noise generated by large value resistors is also significant, so that the resistors used for small signal input stages should be fairly low value, high stability film types, with small currents flowing. Variable resistors must never be used in a low-noise signal stage.

## Voltage gain

The voltage gain of a simple single stage voltage amplifier can be found from a simple rule of thumb. If $V_{bias}$ is the steady d.c. voltage *across the collector load resistor*, then the voltage gain is

$$G = 40 \times V_{bias}$$

For a single stage amplifier, the signal is attenuated both at the input and at the output by the resistance of devices connected to the transistor (microphones, tape heads, other amplifying stages). If the resistance of the signal source is $R_s$ and the resistance of the next stage is $R_{load}$, then the measured gain will be

$$G \times \frac{h_{ie}}{R_s + h_{ie}} \times \frac{R_{load}}{R_{load} + R_c}$$

where $R_c$ is the collector output resistance as shown in *Figure 2.12*, and $G$ is the value of gain given by $40V_{bias}$. This method gives gain values accurately enough for most practical purposes. When precise values of gain are needed, negative feedback circuits (see below) must be used.

*Figure 2.12. The voltage signal equivalent. The voltage gain, G, obtained by the transistor is reduced by the potential divider networks at the input and at the output*

For a multi-stage amplifier, the gains of individual stages are multiplied together, and the attenuations caused by the potential dividing actions of $R_s$ and $R_{load}$ are also multiplied together.

## FETs

Field effect transistors (FETs) are constructed with no junctions in the main current path between the electrodes, which are called drain and source respectively. The path between these contacts, called the channel, may be p-type or n-type silicon, so that both p-channel and n-channel FETs are found. Control of the current flowing in the

*Figure 2.13. MOSFET structure*

channel is achieved by varying the voltage at the third electrode, the gate. In junction FETs, the gate is a contact to a junction formed on the channel and reverse biased in most circuit applications. MOSFETs use a capacitor structure (*Figure 2.13*), so that the gate is completely insulated from the channel. No bias is needed, since the gate is insulated, but care has to be taken to avoid gate breakdown caused by excess voltage. Even electrostatic voltages, generated by contact, can cause damage, so that the gate electrode should be shorted to the source until the MOSFET is wired into circuit. In any circuit application, there must be a resistor connected from gate to source.

The input resistance of either type of FET is very high, and low noise levels can be achieved, even with source resistances as high as $1M\Omega$.

### Negative feedback

Feedback means using a fraction of the output voltage of the amplifier as an input. When the signals at input and output are oppositely phased (mirror-image waveform), then the feedback signal is said to be negative. Negative feedback signals subtract from the input signals to the amplifier, so reducing the overall gain of the amplifier. The effect on gain is as follows:

Let $G$ = Gain of amplifier with no feedback, known as the open loop gain

$n$ = feedback fraction (or loop gain), so that $\dfrac{V_{out}}{n}$ is fed back.

Then the gain of the amplifier when negative feedback is applied is

$$\frac{G}{1 + \dfrac{G}{n}} = \text{the closed loop gain.}$$

For example, if open loop gain, $G = 100$ and $n = 20$ (so that 1/20 of the output voltage is fed back) then the closed loop gain is $\dfrac{100}{1 + \dfrac{100}{20}} = \dfrac{100}{6} =$ 16.7. A very useful approximation is that if the open loop gain $G$ is very much greater than the loop gain, then the closed loop gain (with the negative feedback connected) is simply equal to $n$. This is because $\dfrac{G}{n}$ is large, so that in the expression above it is much greater than unity and the expression is approximately $\dfrac{G}{G/n} = n$.

Negative feedback, in addition to reducing gain, reduces noise signals which originate in the components of the amplifier, and will also

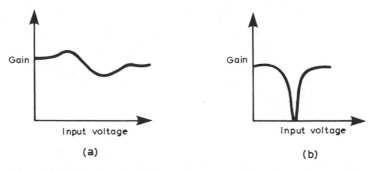

(a)                      (b)

*Figure 2.14. Negative feedback can correct non-linear distortion provided that the gain (before feedback) remains reasonably high (a) over the full range of input voltages. If the gain is zero for any input (b) or is unusually low, then feedback cannot correct the distortion. Cross-over distortion is an example of a fault which causes zero gain.*

reduce distortion provided that the distortion does not cause loss of open-loop gain (*Figure 2.14*). Input and output resistances are also affected in the following way. If the feedback signal shunts the input (*Figure 2.15*) (applied to the same terminal), then input resistance is reduced, often to such an extent that the input terminal is practically at earth potential for signals (a *virtual* earth). If the feedback signal is in series with the input signal (*Figure 2.16*), the input resistance of the amplifier is increased. When the feedback network is connected in parallel with the *output* load (*Figure 2.16*) the effect is to reduce output

*Figure 2.15. A feedback circuit in which the feedback signal is in shunt with the input signal. At the output, the feedback resistor is connected in series with the output load*

*Figure 2.16. A feedback circuit in which the feedback signal is in series with the input signal through the emitter junction of the first transistor. The feedback resistor is also connected in parallel with the output load*

resistance, and when the feedback is connected in series (*Figure 2.15*), the effect is to increase output resistance. The effects on output resistance are generally small compared to the effects on input resistance.

## Heatsinks

A transistor passing a steady (or average) current $I$ and with a steady (or average) voltage $V$ between collector and emitter dissipates a power of $VI$ watts. This electrical power is converted into heat at the collector-base junction, and unless this heat can be removed the temperature of the junction will rise until the junction fails irreversibly. Heat is removed in two stages, by conduction to the case or other metal work of the transistor, and into heatsinks if fitted, then by convection into the air. The temperature of the junction will stabilise when the rate of removing heat, measured in watts, is exactly equal to the electrical power dissipation; this may, however, happen only when the junction temperature is too high for continuous operation. The power dissipation of a transistor is limited therefore mainly by the rate at which heat can be removed.

For practical purposes, the resistance to heat transfer is measured by the quantity called thermal resistance, $\theta$, whose units are $^\circ$C/W. The same measuring units are used for convection from heatsinks as for conduction through the transistor, so that all the figures of thermal resistance from the collector-base junction to the air can be added together as for resistor values in series. The temperature difference between the junction and the air around the heatsink is then found by multiplying the total thermal resistance by the number of watts dissipated

$$T° = θ \times W$$

This latter figure is a temperature *difference,* so that to find the actual junction temperature, the temperature of the air around the heatsink (the ambient temperature) must be added to this figure. An ambient figure of between $30°C$ (for domestic equipment) and $70°C$ (for industrial equipment) may be taken. If this procedure ends with a calculated junction temperature higher than the manufacturer's rated values ($120°C$ to $200°C$ for silicon transistors), then the dissipated power must be reduced, a larger heatsink used, or a water cooled heatsink used. Large power transistors are designed so that the transfer of heat from junction to case is efficient, with a low value of thermal resistance, and the largest thermal resistance in the 'circuit' is that of the heatsink-to-air. Small transistors generally have much higher thermal resistance values, so that the heatsinking is not so effective.

To ensure low thermal resistance, the collector of medium or high-power transistors is connected directly to the case. To prevent unintended short circuits, the heatsink may have to be insulated from other metal work, or the transistor insulated from the heatsink using mica washers. Such washers used with silicone heatsink grease can have thermal resistance values of less than $1°C/W$ and are available from transistor manufacturers or components specialists. The use of mica washers makes it possible to use a metal chassis as heatsink or to mount several transistors on the same heatsink.

$$θ = \frac{250}{H \times D}$$

H and D measured in cm
θ in °C/watt

*Figure 2.17. An approximate guide to the thermal resistance of a metal fin*

The calculation of thermal resistance for heatsinks is not particularly simple, but *Figure 2.17* shows an approximate formula. Measurement of thermal resistance can be carried out by bolting a 25 W wirewound resistor of the metal-cased type to the heatsink. A 2.2 Ω value is suitable and will dissipate 4 W at 3 V and 16.4 W at 6 V. The temperature of the heatsink is measured when conditions have stabilised (no variation of temperature in one minute), and the electrical power figure is divided by the difference between heatsink temperature and ambient (air) temperature. This method is not precise, but gives values which are suitable for practical work.

### Switching circuits

A linear amplifier circuit creates an 'enlarged' copy of a waveform. A pulse (logic) switching circuit charges rapidly from one value of voltage or current to another for a small change of voltage or current at the input. The output waveform need not be similar in shape to the input waveform, but the changes of voltage or current at the output should taken place with only a small time delay (a microsecond or less) after the changes at the input.

The bipolar (or junction) transistor has a good switching action because of its large $G_m$ figure. A useful rule-of-thumb is that the collector current of a transistor will be increased tenfold by an extra 60 mV at the base, provided that the transistor is conducting before the extra base voltage is added, and is not saturated by the extra current. Current switches can thus be easily achieved, and a stage of current amplification can be added if larger current swings or smaller voltage swings are needed (*Figure 2.18*). A voltage switching stage must

*Figure 2.18. Adding a current-amplifying stage to a simple switching transistor*

use some form of load to convert the current changes at the collector into voltage changes. If this load is a resistor, the switch-on of the transistor may be faster than the switch-off. Stray capacitances between the collector and earth are discharged rapidly by the current through the collector at switch-on, but must be charged through the load resistor when the transistor is switched off (*Figure 2.19*). If the rise time of the wave does not need to be short, this can be overcome by using a comparatively low value resistor (1 kΩ or less). An alternative method is shown in *Figure 2.20* using series connected transistors switching in either direction.

For fast switching, the stored charge of transistors can cause problems. During the time when the transistor is conducting, the emitter is injecting charges into the base region. These charges cannot disappear instantly when the base bias is reversed, so that the transistor will conduct momentarily in the reverse direction. As a result, the

*Figure 2.19. Charging and discharging stray capacitances. When the transistor conducts (a) the stray capacitance is rapidly discharged, and the voltage drop at the collector is sharp. When the transistor cuts off (b), the stray capacitance is recharged through the load resistor, causing a slower voltage rise*

circuit of *Figure 2.20* can suffer from excessive dissipation at high switching speeds, since for short intervals, both transistors will be conducting. Manufacturers of switching transistors at one time quoted figures of stored charge, $Q$ in units of picocoulombs (pC) or nano-coulombs (nC), but nowadays, generally quote the more useful turn-on

*Figure 2.20. Using a two-transistor output circuit so that the switching is equally rapid in both directions. This type of output stage is used in TTL digital i.c.s*

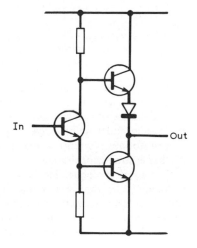

and turn-off times, in nanoseconds (ns) under specified conditions. If only stored charge figures are given, an approximate value for turn-off time can be obtained from the equation

$$t = \frac{Q}{I}$$

where $t$ is turn-off time in nanoseconds ($10^{-9}$s), $Q$ is stored charge in pC and $I$ is the current in mA which is to be switched off. Transistor switch-off times are improved by reverse-biasing the base, but some care has to be taken not to exceed the reverse voltage limits, since the base-emitter junction will break down at moderate voltages.

A considerable improvement in switch-off times is also obtained if the transistor is not allowed to saturate during its switch-on period; this has to be done by clamping the base voltage and is not easy because of the considerable variation of switch-on voltage between one transistor and another. The fastest switching times are achieved by 'current switching' circuits, in which the transistor is never saturated nor cut-off.

(a)                              (b)

*Figure 2.21. Two common switching circuit tricks. (a) use of a base-compensation capacitor, (b) using a complementary output circuit with no load resistor*

Some circuits commonly used for switching circuits are shown in *Figure 2.21*, where *(a)* shows the use of a time constant $RC$ in series with the base of the transistor; $C$ should be adjusted for the best shape of leading and trailing edges. *Figure 2.21b* shows the familiar double-emitter-follower circuit which uses transistors both to charge and to discharge stray capacitance.

Gating of analogue signals is an action similar to that of pulse (logic) switching, but the switch may be a series component rather than a shunt component, with the added reduction that it should not distort the analogue signal while in the ON state. Diodes, bipolar transistors, and FETs can all be used in such gating circuits. *Figure 2.22* shows a

*Figure 2.22. A diode-bridge gate circuit. In the off state, diodes $D_1$ and $D_6$ conduct, so that no current flows through $D_2$, $D_3$, $D_4$, $D_5$. When the gate is switched on by symmetrical pulses, $D_1$ and $D_6$ shut off, allowing the other diodes to conduct, so that the input signal can reach the output. Using symmetrical switching signals ensures that very little of the switching waveform appears in the output signal*

diode bridge gate. When current flows through the diodes, assuming that there is a large resistance between point A and earth and between point B and earth, then there is a low resistance path C — D for signals in either direction. If the diodes are well matched, the d.c. level at D should be identical to that at C. Such a voltage difference is called an offset and is undesirable. When current ceases to flow, the diodes

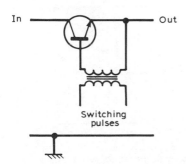

*Figure 2.23. A gate circuit using a single transistor*

become non-conducting, the gate closes. Offset voltages may be around 10 mV (comparatively high), but very high operating speeds are possible when switching diodes are used.

Bipolar transistors can be used as gates, with the offset voltage between collector and emitter being lowest (less than 2 nV) for a saturated transistor when the normal collector and emitter terminals are *reversed*. As a series switch, however, the transistor suffers from the requirement of a switching pulse applied between base and emitter, so that a transformer must be used to apply the switching voltage if a single transistor is to be used (*Figure 2.23*). A very common gating circuit is the long-tailed pair of *Figure 2.24*, which is, however, useful

Figure 2.24. The long-tailed pair gate. When $Tr_2$ is switched off $Tr_1$ is normally biased by $R_1$, $R_2$, and acts as an inverting amplifier. When $Tr_2$ is switched on, with its base voltage several volts higher than the normal bias voltage of $Tr_1$, then $Tr_1$ is biased off

only when the offset voltages ($v_{be}$) and the voltage change caused by switching are unimportant.

FETs may be used as shunt or series switches (*Figure 2.25*). The shunt switch is considerably easier to drive because the source terminal is earthed. Offset voltages of less than 10 $\mu$V are obtainable, and practically no drive current to the gate is needed. The disadvantage of the FET is that its resistance when switched ON is much higher (up to 1 k$\Omega$) than that of a bipolar transistor.

Figure 2.25. Using FETs as switches: (a) shunt, (b) series

## Other switching devices

Unijunctions have two base contacts and an emitter contact, forming a device with a single junction which does not conduct until the voltage between the emitter and base contact 1 (*Figure 2.26*) reaches a specified level. At this level, the whole device becomes conductive. The unijunction

B2 Base 2

Emitter

B1 Base 1

*Figure 2.26. Unijunction symbol*

is used to generate short pulses, using circuits such as that of *Figure 2.27*. The frequencies of operation of this circuit are not noticeably affected by supply voltage changes, since the unijunction fires (becomes conductive) at a definite fraction of the supply voltage. The intrinsic stand-off ratio, $n$, is defined as

$$\frac{\text{firing voltage } (e - b_1)}{\text{supply voltage } (b_2 - b_1)}$$

and has values ranging typically from 0.5 to 0.86. Pulse repetition rates up to 1 MHz are obtainable.

*Figure 2.27. A unijunction oscillator. $R_2$, $R_3$ are about 100 ohms each, and the frequency of oscillation is determined by the time constant RC*

Programmable unijunction transistors (PUTs) have three terminals, one of which is used to set the value of intrinsic stand-off ratio, $n$, by connection to a potential divider (*Figure 2.28*). Firing will occur at the programmed voltage; the frequency range is generally up to 10 kHz.

Thyristors are controlled silicon diodes which are switched into conduction by a brief pulse or a steady voltage at the gate terminal. Voltage of 0.8 to 1.5 V and currents of a few $\mu$A up to 30 mA are needed at the gate, according to the current rating of the thyristor.

*Figure 2.28. A programmable unijunction transistor (PUT). The firing voltage between anode and cathode is selected by the voltage applied to the third electrode*

*Figure 2.29. A capacitor turn-off circuit for a thyristor. When the switch is momentarily closed, the sudden voltage drop at A will cause an equal drop at X, turning off the thyristor until it is triggered again*

*Figure 2.30. A.C. thyristor circuits. (a) Basic half-wave a.c. relay circuit. (b) A full-wave relay circuit. (c) Basic phase-control circuit. In the half-cycle during which the thyristor can conduct, the gate is activated only when the voltage at A has risen enough to cause the trigger diode $D_1$ to conduct. The time in the cycle at which conduction starts is controlled by the setting of the variable resistor*

The thyristor ceases to conduct only when the voltage between anode and cathode falls to a low value (about 0.2 V) or when the current between anode and cathode becomes very low (less than 1 mA). D.C. switching circuits need some form of capacitor discharge circuit (*Figure 2.29*) to switch off the load. A.C. switching circuits, using a.c. or full wave rectified waveforms, are switched off by the waveform itself on each cycle. A few typical a.c. thyristor circuits are shown in *Figure 2.30*. Note that the gate signal may have to be applied through a pulse transformer, particularly when the thyristor switches mains currents, to avoid connecting the firing circuits to the gate.

Triacs are two-way thyristors whose terminals are labelled MT1, MT2 and gate. For reliable firing, the pulse at the gate should be of the same polarity as MT2 (some circuits are shown in *Figure 2.31*).

*Figure 2.31. Triac circuits. (a) Basic full-wave relay circuit. (b) Power regulator circuit, using a diac trigger diode, and radio interference suppression circuit across the triac*

Firing pulses for thyristors and triacs can be obtained from uni-junctions or from other types of trigger device such as diacs, silicon bidirectional switches, four-layer diodes or silicon unidirectional switches. The diac, or bidirectional trigger diode, is non-conductive in either direction until its breakdown voltage is exceeded, after which the device conducts readily until the voltage across its terminals (either polarity) is low. Firing voltages of 20 to 36 V are typical, and the 'breakback' voltage (at which the device ceases to conduct) is typically 6 V. Brief peak currents of 2 A are possible. The silicon bidirectional switch also uses a gate electrode, but operates with one polarity only. Four-layer diodes have lower firing and breakback voltages than diodes, but essentially similar characteristics.

The silicon controlled switch (SCS) is a useful device with four electrodes which can be used, according to connections, either as a programmable unijunction or as a low-power thyristor. The connections are referred to as anode, cathode, gate-anode and gate-cathode. If the gate-cathode is used together with the anode and cathode, thyristor operation (at low currents) is obtained; if the gate-anode is used, the

### Table 2.1  PRO-ELECTRON CODING

The first letter indicates the semiconductor material used:

| | |
|---|---|
| A | Germanium |
| B | Silicon |
| C | Gallium arsenide and similar compounds |
| D | Indium antimonide and similar compounds |
| R | Cadmium sulphide and similar compounds |

The second letter indicates the application of the device:

| | |
|---|---|
| A | Detector diode, high speed diode, mixer diode |
| B | Variable capacitance (varicap) diode |
| C | A.F. (not power) transistor |
| D | A.F. power transistor |
| E | Tunnel diode |
| F | R.F. (not power) transistor |
| G | Miscellaneous |
| L | R.F. power transistor |
| N | Photocoupler |
| P | Radiation detector (photodiode, phototransistor, etc.) |
| Q | Radiation generator (LED etc.) |
| R | Control and switching device (such as thyristor) |
| S | Switching transistor, low power |
| T | Control and switching device (such as a triac) |
| U | Switching transistor, high power |
| X | Multiplier diode (varactor or step diode) |
| Y | Rectifier, booster or efficiency diode |
| Z | Voltage reference (zener), regulator or transient suppressor diode. |

The remainder of the code is a serial number. For consumer applications, such as radio, TV, hi-fi, this has three figures. For industrial and telecommunications use, a letter W, X, Y or Z and two figures are used.

device behaves as a PUT. The unused electrode is generally left open circuit.

*Table 2.1* shows the European Pro-Electron coding used for semiconductors. The US (JEDEC) 1N and 2N numbers are registration numbers only (this also applies to the Japanese 2SA, 2SB etc. system), and the function of a semiconductor cannot be guessed from the number.

# Chapter 3

# Discrete Component Circuits

This Chapter illustrates a selection of well established circuits and data, and comments are reduced to a minimum so as to include the greatest possible number of useful circuits. The common-emitter and a few other amplifier circuits have already been dealt with in Chapter 2.

Where several different types of circuits are shown (as for oscillators) practical considerations may dictate the choice of design. For example,

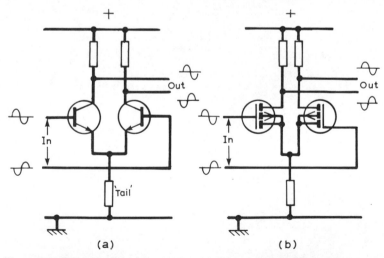

*Figure 3.1. The long-tailed pair circuit using (a) bipolar transistors, (b) p-channel MOSFETS. Balanced input signals, as shown, are amplified, but unbalanced signals (in the same phase at each input) are attenuated*

a Hartley oscillator uses a tapped coil, but the arrangements for frequency variation may be more convenient than those for a Colpitts oscillator, which uses a capacitive tapping. Some crystal oscillator circuits are not always self-starting, particularly with 'difficult' crystals. For this reason, as many variations on basic circuits have been shown as is feasible within the space.

The long-tailed pair, shown both in bipolar and in FET form in *Figure 3.1* is the most versatile of all transistor circuits, which is why it is so extensively used for linear i.c.s. A common-mode signal is a signal applied in the same phase to both bases or gates. Any amplification of such a common mode signal can only be caused by lack of balance between the transistors, so that this value of gain is low. The difference signal is amplified with a considerably greater gain. The long-tailed pair is most effective when used as a balanced amplifier, with balanced input and output, but single-ended inputs or outputs can be provided, as shown in *Figure 3.2a* and *b*.

Figure 3.2. Single-ended inputs and outputs on a long-tailed pair circuit. The second input is earthed to signals — no bias arrangements have been shown. (a) Bipolar transistors, (b) p-channel MOSFETs

*Figure 3.3* illustrates a magnetic pickup preamplifier circuit. The problem here is to apply the frequency correction (equalisation) needed for disc replay. Discs are recorded to the RIAA standard (see BS 1928: 1965), in which bass frequencies are attenuated (to prevent excessive cutter movement) and treble frequencies are boosted (to have an amplitude well above that of surface noise). This deliberate distortion must be corrected at playback by *CR* networks; to achieve this correctly, three time constants, 75 $\mu$s, 318 $\mu$s and 3180 $\mu$s, are used. The 75 $\mu$s time constant is of a treble cut filter, taking effect (that is, with its turnover point) at 2.12 kHz; the 318 $\mu$s kHz is a bass boost starting at 500 Hz, and the 3180 $\mu$s time constant is a final stage of bass cut at 50 Hz and under.

*Figure 3.3. An input stage for magnetic pickup cartridges. This stage combines high gain with equalisation, and a fairly high voltage supply (40 V or more) is needed to avoid overloading on transients*

The output from most magnetic cartridges, other than the moving-coil type, is around 5 mV at the standard conditions of 5 cm/s stylus velocity and 1 kHz signal. This corresponds to a signal amplitude of close to the minimum most amplifiers need to give full output for 2.5 mV input at full volume setting. The preamplifier should present an input resistance of around 50 k$\Omega$ and should be capable of accepting considerable overloads at the input, 50 mV or more, without noticeable distortion. The trend at the time of writing is away from the feedback-loop type of equalisation, and towards an equalising network of purely passive type applied after a 'flat' preamplifier stage — this is held to be advantageous because of the effects of transients on feedback amplifiers, particularly when momentarily overloaded.

*Figure 3.4* shows some tape/cassette input circuits. Once again, equalisation is needed, but the time constants are not quite so universally agreed; they appear to change almost annually as new tape materials and new types of tape head construction appear. In addition to these 'standard' corrections, individual tape decks may need further corrections, a multiplex filter may be included to remove f.m. stereo subcarrier

Figure 3.4. Tape equalisation and input stages. (a) One form of feedback network which can be used for reel-to-reel tape equalisation. (b) A cassette recorder input stage, using rather different time constants in the equalisation networks

signals, and noise-cancelling circuits, such as the Dolby circuits, may be used. At the last count, the equalisation frequencies being used on replay were 3180 $\mu$s for all tapes and either 70 $\mu$s or 120 $\mu$s for chrome and ferric tapes respectively; ferrichrome and pure iron particle tapes are replayed at 70 $\mu$s. The equalisation needed for recording amplifiers is too specialised to include here partly because recording equalisation time constants depend much more on individual needs.

*Figute 3.5. A Baxandall type of tone control circuit. This circuit is normally located between two voltage amplifier transistors which provide the necessary gain*

*Figure 3.5* shows a version of the Baxandall tone control circuit, which is virtually the standard method of tone control used nowadays. There is very little interaction between the treble and the bass controls, low distortion, and a good range of control; about 20 dB of boost or cut. The Baxandall circuit is usually located between bipolar transistors, but several designs claim significant improvements by using a FET at the output of the control stage.

*Figure 3.6* deals with active filters. These designs use only resistors and capacitors, together with semiconductors, and are considerably simpler to design than *LC* filters. Low-pass, high-pass, bandpass and notch filters are illustrated in *Figure 3.6*. The filters generally have a slope of 12 dB per octave (meaning that the response changes by 12 dB for each doubling or halving of frequency).

*Figure 3.7* shows a typical input preamplifier stage for a moving-coil microphone. The particular features here are low noise operation, matching a fairly low input resistance, high gain, and hum rejection. The low output and low resistance of the moving-coil microphone requires the use of a microphone transformers. If a balanced layout is possible, hum pickup can be greatly reduced.

Figure 3.6. Active filter circuits. (a) Low pass, (b) High pass. These two types are Sallen & Key filters, named after the inventors. Circuit (c) is a bandpass filter using the Wien bridge network and an inverting amplifier. The gain of the amplifier determines the effective Q of the circuit, so that higher gain causes narrower bandwidth. Circuit (d) is a passive twin-T notch filter. The selectivity can be improved by using positive feedback

*Figure 3.7. A moving-coil microphone input circuit. A suitable
input transformer is essential if high-quality results are expected.
The transformer should preferably be supplied by the makers of
the microphone*

The next three sets of circuits deal with audio output stages. Class A
stages are those in which the transistor(s) are always biased on and
never saturated (bottomed). A Class A stage may use a single transistor
(a single-ended stage) or two transistors which share the current in some
way (a push-pull stage), but the efficiency is low. % Efficiency is
defined as

$$\frac{\text{power dissipated in the load} \times 100}{\text{total power dissipated in the output stage}}$$

and is always less than 50% for Class A operation.

A Class A stage should pass the same current when no signal is applied
as when maximum signal is applied. Because of this, the dissipation is
large, so that large-area heatsinks are needed for the output transistors.

*Figure 3.8. A Class A
single-ended output stage.
Good heatsinking is es-
sential*

Class B audio operation uses two (or more) transistors biased so that one conducts on one half of the waveform and the other on the remaining half. Some bias must be applied to avoid 'crossover distortion' due to the range of base-emitter voltage for which neither transistor would conduct in the absence of bias. Class B audio stages can have efficiency figures as high as 75%, though at the expense of rather higher distortion than a Class A stage using the same layout. The higher efficiency enables greater output power to be obtained with smaller heatsinks, and the use of negative feedback can, with careful design, reduce distortion to negligible levels.

*Figure 3.8* shows a Class A single-ended power output stage, suitable for general-purpose use such as car radio operation. *Figure 3.9* shows the totem-pole or single-ended push-pull circuit, which can be used for either Class A or Class B operation according to the bias level. This version uses complementary symmetry — the output transistors are

*Figure 3.9. A single-ended push-pull (totem-pole) Class B output stage, using complementary power transistors. $VR_1$ sets the voltage at point X to half of the supply voltage, $VR_2$ sets the quiescent (no signal) current through the output transistors. $C_1$ is a 'bootstrap' capacitor which feeds back in-phase signals to point Y, increasing input impedance. Oscillation is avoided because the gain of $Tr_2$ is less than 1*

pnp and npn types. When complementary output transistors cannot be obtained, a pseudo-complementary circuit, such as that of *Figure 3.10*, can be used, though this is not truly symmetrical.

*Figures 3.11, 3.12* and *3.13* illustrate some of the circuits used for wideband voltage amplification. *Figure 3.11* deals with methods of frequency compensation using inductors or capacitors to compensate for the shunting effect of stray capacitances. *Figure 3.12* shows a circuit which uses feedback to reduce the gain and so extend the flat portion

*Figure 3.10. A quasi-complementary output stage. The low power complementary transistors $Tr_1$, $Tr_2$ drive the high-power output pair, $Tr_3$, $TR_4$, which are not complementary types. The circuit is not symmetrical, and can cause considerable distortion when overdriven*

of the frequency range — this is a useful basic circuit for video frequencies. *Figure 3.13* shows a cascode amplifier, a type of construction which was widely used in the days of valves, but which has been strangely neglected in transistor circuits. The advantage of the cascode is stability, because there is practically no feedback from output to input and high gain over a large bandwidth. FET cascodes and combinations of FET and bipolar transistors can also be used.

Figure 3.11. Frequency compensation for wideband amplifiers. (a) Capacitive compensation. The value of $C_2$ is chosen so that $R_4$ is progressively decoupled at high frequencies. As a rough guide, $C_sR_3$ should equal $C_2R_4$. (b) Inductive shunt compensation. The value of L is chosen so as to resonate with the input capacitance of the transistor at a frequency above that of the uncompensated 3 dB point. These compensation methods are useful, but cannot compensate for low gain caused by an unsuitable transistor type. Transistors capable of amplification at high frequencies must be used in these circuits

Figure 3.12. A feedback pair circuit which is capable of wideband amplification when suitable transistors are used. Bandwidths of 5 MHz or more are obtainable

*Figure 3.13. The cascode circuit, using, in this example, bipolar transistors. This is a exceptionally stable circuit, because of the isolation between input and output, and has high gain and a large bandwidth*

The circuits of *Figures 3.14* to *3.16* are of sine wave oscillators which operate at radio frequencies. The Hartley type of oscillator (*Figure 3.14*) uses a tapped coil; the Colpitts type (*Figure 3.15*) uses a capacitor tap. Though these are not the only r.f. oscillator circuits, they are the circuits most commonly used for variable frequency oscillators. *Figure 3.16* illustrates some crystal controlled oscillator circuits. The frequency of the output need not be the fundamental crystal frequency, since most crystals will oscillate at higher harmonics (overtones) and harmonics can be selected at the output. Frequency multiplier stages (see *Figure 3.26*) can then be used to obtain still higher frequencies.

*Figure 3.14. The Hartley oscillator. The resonant circuit is $L_1C_1$, and the value of $C_2$ should be chosen so that the amount of positive feedback is not excessive, since this causes a distorted waveform. $R_1$ should be chosen so that the transistor is just drawing current when $C_1$ is short-circuited*

*Figure 3.15. The Colpitts oscillator. The tapping is provided by $C_2$, $C_3$, and it is the series combination of these two capacitors which tunes $L_1$. As a rough guide, the value of $C_3$ should be about ten times the value of $C_2$ to avoid over-driving. $R_1$ is chosen to give about 1 mA of collector current. $C_1$ must not be omitted, because oscillation is impossible (because of negative feedback to the base) unless the base is decoupled*

For low frequencies, oscillators such as the Wien bridge (*Figure 3.17*) or twin-T types are extensively used. Usable frequency ranges are from 1 Hz, or lower, to around 1 MHz.

Untuned or aperiodic oscillators are important as generators of square and pulse waveforms. *Figure 3.18* shows the familiar multivibrator (astable) together with modifications which improve the shape of the waveform. The less familiar serial multivibrator is shown in *Figure 3.19;* this circuit is a useful source of narrow pulses. When a pulse of a determined, or variable, width is required from any input (trigger)

(a)                    (b)

*Figure 3.16. Two versions of a Colpitts crystal oscillator, showing alternative positions for the cystal* pulse, a monostable circuit must be used. *Figure 3.20* shows a mono-stable circuit, with a block diagram to illustrate how a combination of astable and monostable can form a useful pulse generator.

*Figure 3.21* shows the basic bistable circuit, now rather a rarity in the discrete form thanks to the low price of i.c. versions. The Schmitt trigger is illustrated in *Figure 3.22*; its utility is as a comparator and

(a)

(b)

Figure 3.17. Low-frequency oscillators. (a) Wien bridge, (b) twin-T. The Wien bridge circuit uses a thermistor to keep the amplitude of the output signal constant. $R_1$, $R_2$ may be a ganged variable if a variable frequency output is needed

$$T_1 = 0.7\, C_1 R_2$$
$$T_2 = 0.7\, C_2 R_3$$
$$f = \frac{1}{T_1 + T_2}$$

*Figure 3.18. The astable multivibrator. The frequency of operation is given by the formula shown. The diodes $D_2$, $D_3$ prevent breakdown of the base-emitter junctions of the transistors when the transistors are turned off, and $D_1$ isolates the collector of $Tr_2$ from $C_2$ when $Tr_2$ switches off. In this way, a fast-rising waveform can be obtained*

trigger stage which gives a sharply changing output from a slowly changing input. The hysteresis (voltage difference between the switching points) is a particularly valuable feature of this circuit. A circuit with hysteresis will switch positively in each direction with no tendency to 'flutter' or oscillation, so that Schmitt trigger circuits are used extensively where electronic sensors have replaced purely mechanical devices such as thermostats.

$$T = 0.7\, C_1 R_1$$

*Figure 3.19. The serial astable. Only one time constant is needed, and the outputs are a sawtooth and a pulse as shown. Usually $R_3 = R_4$; values of around 10 kΩ are usual*

Radio-frequency circuits are represented here by only a few general examples, because the circuits and design methods that have to be used are fairly specialised, particularly for transmission, and the reader who wishes more information on purely r.f. circuits is referred to the excellent amateur radio publications. *Figure 3.23* shows an a.m./f.m. i.f. amplifier for 470 kHz and 10.7 MHz such as would be used in a.m./ f.m. receivers. This design uses a common emitter amplifier, since the

(a)

(b)

*Figure 3.20. The monostable (a). The pulse width of the output pulse is determined by the time constant CR. The block diagram (b) shows how an astable (to determine frequency) and a monostable (to determine pulse width) can be combined to form a precise pulse generator*

operating frequency is well below the turnover frequency, $f_r$ (at which gain is unity) for the transistor. For v.h.f. use, common base stages (*Figure 3.24*) and dual-gate FETs (*Figure 3.25*) are extensively used. *Figure 3.26* shows frequency multiplier and intermediate stages for transmitters, and *Figure 3.27* a selection of low-pair output (power

Figure 3.21. The bistable, or flip-flop. The output changes state (high to low or low to high) at each complete input pulse

(a)

(b)

Figure 3.22. The Schmitt trigger circuit. (a) This is a switching circuit whose output is always flat-topped and steep-sided whatever the input waveform. The characteristic (b) shows hysteresis — a difference between the switching voltages depending on the direction of change of the input voltage

Figure 3.23. An i.f. amplifier typical of a.m./f.m. receivers. The frequency difference between the i.f.s is so great that no special filtering is needed, but the 10.7 MHz transformers must be located close to their transistors

Figure 3.24. A typical f.m. 'front-end', using a common-base r.f. amplifier and a common-base oscillator. All transistors will operate at higher frequencies in the common-base connection than in the common-emitter connection

A.F.C. in

A.G.C. (2) in

A.G.C. (1) in

Figure 3.25. An f.m. front end of a high-quality tuner (Armstrong), using dual-gate MOSFETs. Using both gates enables the r.f. a.g.c. to be completely isolated from the signal input, and the mixer signal input to be isolated from the oscillator input amplifier or p.a.) stages. Transmitters which use variable frequency oscillators (v.f.o.) will require broadband output stages as distinct from sharply tuned stages, and this precludes the use of Class C amplifiers (in which the transistor conducts only on signal peaks). Without a sharply tuned, high $Q$, load, Class C operation introduces too much distortion (causing unwanted harmonics) and so Class B is preferable.

(a)                                    (b)

*Figure 3.26. Frequency multipliers for transmitters. (a) A single transistor multiplier for even or odd multiples, (b) a push-pull multiplier for odd multiples. In each type of circuit, the output is tuned to a frequency which is a multiple of the input frequency. Other techniques not shown here include push-push, in which two transistors have antiphase signals at their collectors, but both feed the same output at the collectors. This circuit is used for even multiples. Varactor diodes are also extensively used as multipliers at low power levels*

To carry information by radio or by digital signals requires some form of modulation and demodulation. For radio use amplitude modulation and frequency modulation are the most common techniques. Straightforward amplitude modulation produces two sidebands, with only one third of the total power in the sidebands, so that double sideband a.m. is used virtually only for medium and long wave broadcasting. Short wave communications use various forms of single sideband or suppressed carrier a.m. systems; v.h.f. radio broadcasting uses wideband f.m. and other v.h.f. communications use narrow-band f.m.

*Figure 3.28* shows two simple modulator circuits, excluding specialised types. Carrier suppression can be achieved by balanced modulators in which the bridge circuit enables the carrier frequency to be balanced out while leaving sideband frequencies unaffected.

Sideband removal can be achieved by crystal filters, a fairly straight-forward technique which is applicable only if the transmitting frequency is fixed, or by a phase-shift modulator which makes use of the phase shift which occurs during modulation. Frequency modulation, unlike amplitude modulation, is carried out on the oscillator itself, so requiring

*Figure 3.27. Power amplifiers for transistor transmitters. (a) A Class C single transistor p.a. stage, (b) a Class B design, necessary for single-sideband transmitters. Tuning inductors have been omitted for clarity. In both circuits some decoupling capacitors have not been shown — complete decoupling is essential. At the higher frequencies, circuit layout is critical, and the circuit diagram becomes less important than the physical layout*

reasonably linear operation of the stages following the oscillator. *Figure 3.29* illustrates some types of discrete component demodulators.

Pulse modulation systems are used extensively in applications ranging from data processing to radar. Pulse amplitude modulation and frequency modulation is essentially similar in nature to a.m. and f.m.

of sine waves, and will not be considered here. Forms of modulation peculiar to pulse operation are pulse width modulation (p.w.m.), pulse position modulation (p.p.m.) and pulse-code modulation (p.c.m.). A technique which is not a pulse modulation system but which is

*Figure 3.28. Two simple modulator circuits. (a) A collector modulated stage for an a.m. transmitter, (b) a varactor diode f.m. modulator*

extensively used for coding slow pulse information is frequency shift keying (f.s.k.) in which the high (logic 1) and low (logic 0) voltages of a pulse are represented by different audio frequencies.

*Figure 3.30* is concerned with optical circuits, including LED devices and light detectors.

The circuits of *Figure 3.31* deal with power supply units. *Figure 3.31* shows the no-load voltage output, and the relationship between d.c. load voltage, minimum voltage, and a.c. voltage at the transformer. Only capacitive input circuits have been shown, since choke-input filters are by now rather rare. The relationship between the size of the reservoir capacitor and the peak-to-peak ripple voltage is given approximately by

$$V = \frac{I_{dc} \times t}{C}$$

with $I_{dc}$ equal to load current (amperes), $t$ in seconds the time between voltage peaks and $C$ the reservoir capacitance in farads.

A more convenient set of units is $I_{dc}$ in mA, $t$ in ms, and $C$ in $\mu$F, using the formula unchanged. $V$ is then the peak-to-peak ripple voltage in volts.

(a)

(b)

*Figure 3.29. Demodulators. The a.m. demodulator (a) uses a single diode. The time constant of $C_1$ with $R_1 + R_2$ must be long compared with the time of a carrier wave, but short compared with the time of the highest-frequency audio wave. The f.m. demodulator (b) is a ratio detector*

*Figure 3.30. Optoelectronic circuits. (a) Driving a single LED — a current-limiting resistor must be used. (b) When an LED is operated from a.c., a diode must always be included to protect the LED from reverse voltage. (c) Optocoupler, used to couple signals at very different d.c. levels. This is useful for triac firing, or for modulating the grid of a c.r.t., since d.c. signals can be transferred, which is not possible using a transformer*

All power supplies which use the simple transformer-rectifier-capacitor circuit will provide an unstabilised output, meaning that the output voltage will be affected by fluctuations in the mains voltage level and also by changes in the current drawn by the load. The internal resistance of the power supply unit causes the second effect and can be the reason for instability in amplifier circuits, or of misfiring of pulse circuits. A stabiliser circuit provides an output which, ideally, remains constant despite any reasonable fluctuation in the mains voltage and

If $\tilde{V}$ is r.m.s. input voltage:
$V_{dc}$, no load $= 1\cdot4\,\tilde{V}$
Diode peak reverse voltage $= 2\cdot8\,\tilde{V}$
Minimum voltage, full load current $= 0\cdot44\tilde{V}$
Ripple at line frequency (50 Hz)

Note $\tilde{V}$ for whole of secondary winding
$V_{dc}$, no load $= 0\cdot7\,\tilde{V}$
Diode peak reverse voltage $= 1\cdot4\tilde{V}$
Minimum voltage, full load current $= 0\cdot44\tilde{V}$
Ripple at double line frequency (100 Hz)

$V_{dc}$, no load $= \tilde{V}$
Diode peak reverse voltage $= 1\cdot4\,\tilde{V}$

Minimum voltage, full load current $= 0\cdot44\tilde{V}$
Ripple at double line frequency (100 Hz)

*Figure 3.31. Rectifier circuits in detail*

has zero internal resistance so that the output voltage is unaffected by the load current.

Stabilisation is achieved by feeding into the stabiliser circuit a voltage which is higher than the planned output voltage even at the worst combination of circumstances — low mains voltage and maximum load current. The stabiliser then controls the voltage difference between input and output so that the output voltage is steady.

Design data: Allow 2mA minimum current through zener diode

Maximum current $=( I_{load} + 2)$mA

Diode dissipation, max. $= V_0 ( I_{load}+2)$mW

Resistor dissipation, max $=$
$(V_{in} - V_0)(I_{load} +2)$ mW

*Figure 3.32. A simple zener-diode stabiliser*

*Figure 3.32* shows a simple zener diode stabiliser suitable for small scale circuits taking only a few milliamps. This is a shunt stabilising circuit, so called because the stabiliser (the zener diode) is in parallel (shunt) with the load. The value of the resistor $R$ is such that there will be a 'holding' current of 2 mA flowing into the zener diode even at the lowest input voltage and maximum signal current. The circuit of *Figure 3.33* (sometimes known as the 'amplified zener') is a shunt stabiliser which does not depend on dissipating power in the zener when the load current drops.

*Figure 3.33. An 'amplified-zener' or shunt-regulator circuit. The transistor dissipation is greatest when the load current is least*

*Figure 3.34. A simple series stabiliser. The transistor dissipation is greatest when the load current is maximum*

*Figure 3.34* shows a simple series stabiliser, using a zener diode to set the voltage at the base of an emitter follower. A more elaborate negative feedback circuit is shown in *Figure 3.35,* with provision for altering the stabilised voltage. The circuit of *Figure 3.36* also provides automatic shut-off (with the shut-off current alterable by setting the

*Figure 3.35. A variable-voltage series stabilised supply circuit. $Tr_1$ is usually a 2N3055, and $Tr_2$ a lower power general purpose transistor. The most suitable i.c. is the LM3900, because it continues to amplify even when the output voltage is close to either supply voltage, unlike the 741*

*Figure. 3.36. A fixed-voltage stabiliser with over-current protection, limiting the current to an amount which is set by the potentiometer*

100 Ω potentiometer) when excessive load current flows. More elaborate circuits are not considered here because of the extensive use of i.c. regulators (see Chapter 4).

# Chapter 4

# Linear I.C.s

Linear i.c.s are single-chip arrangements of amplifier circuits that are intended to be biased and operated in a linear way. This definition is usually extended to include i.c.s which have a comparatively slow switching action, such as the 555 timer.

The most important class of linear amplifier i.c. is the operational amplifier which features high gain, high input resistance, low output resistance and a narrow bandwidth extending to d.c. Such amplifiers are almost invariably used in negative feedback circuits, and make use of a balanced form of internal circuit (*Figure 3.1*) so that power supply hum and noise picked up by stray capacitance are both discriminated against.

The 741 is typical of operational amplifiers generally, so that the design methods, circuits and bias arrangements which are used for this i.c. can be used, with small modifications, for other types. Referring to the pinout diagram and symbol of *Figure 4.1*, the 741 uses two inputs marked + and −. These signs refer to the phase of the output signal relative to each input, so that feedback directly from the output to the + input is positive, and feedback directly from the output to the − input is negative.

The circuit arrangement of the 741 is such that, using balanced power supplies, the d.c. level at the output ought to be at zero volts when both inputs are connected to zero volts. This does not generally happen because of slight differences in internal components, so that an *input offset voltage* is needed to restore the output to zero voltage. Alternatively, the offset can be balanced out by a potentiometer connected as shown in *Figure 4.2*. Once set in this way so that the

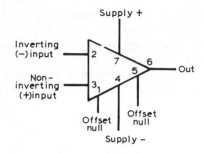

Pin numbers
as viewed
from above

Maximum ratings

Positive supply +18V
Negative supply − 18V
Differential input voltage ± 30V
Common-mode input voltage ± 15 V
Load current   10 mA
Operating temperature  70°C
Storage temperature   140°C
Dissipation  400 mW

*Figure 4.1. 741 operational amplifier outline, with pin numbering
(a) and the connections (b). The offset-null pins are used only for
d.c. amplifier applications*

*Figure 4.2. Using an offset-null control. With the
inputs both earthed (balanced power supplies) and
a voltmeter connected to the output (dotted lines),
the 10 KΩ potentiometer is adjusted so that the
output voltage is zero*

output is at zero volts (with the inputs earthed), the output voltage will then slowly change (drift). The drift may be caused by temperature changes, by supply voltage changes, or simply by old age. Drift is a problem which mainly affects high-gain d.c. coupled amplifiers and long time-constant integrators; a.c. amplifier circuits, and circuits which can use d.c. feedback bias are not affected by drift.

## Bias Methods

For linear amplification, both inputs must be biased to a voltage which lies approximately halfway between the supply voltages. The output voltage can then be set to the same value by:

(a)    making use of an offset-balancing potentiometer, or,

(b)    connecting the output to the (−) input through a resistor, so making use of d.c. feedback.

Method (a) is seldom used, and the use of d.c. feedback is closely tied up with the use of a.c. feedback; the two will be considered together.

The power supply may be of the balanced type, such as the ±15 V supply, or unbalanced, provided that the bias voltage of input and output is set about midway between the limits (+15 and −15, or +V and 0) of supply voltages. Bias voltages should not be set within three volts of supply voltage limits, so that when a ±15 V supply is used, the input or output voltages should not exceed +12 V or −12 V. This limitation applies to bias (steady) voltage or to instantaneous voltages. If a single-ended 24 V power supply is used, the input and output voltages should not fall below 3 V nor rise above 21 V. Beyond these limits, the amplifying action may suddenly collapse because there is not sufficient bias internally.

## Basic circuits

*Figure 4.3* shows the circuits for an inverting amplifier, using either balanced or unbalanced power supplies. The d.c. bias conditions are set by connecting the (+) input to mid-voltage (which is earth voltage when balanced power supplies are used) and using 100% d.c. feedback from the output to the (−) input. The gain is given by:

$$G = R_1/R_2$$

Note that a capacitor $C_1$ is needed when a single-ended power supply is used to prevent the d.c. bias voltage from being divided down in the same ratio as the a.c. bias. When balanced power supplies are used, direct coupling is possible provided that the signal source is at zero d.c. volts.

The input resistance for these circuits is simply the value of resistor $R_2$, since the effect of the feedback is to make the input resistance at the (−) input almost zero; this point is referred to as a 'virtual earth' for signals. The output resistance is typically about 150 ohms.

Typically, $R_1 = 220k$
$R_2 = 22k$
$R_3 = 22k$
Gain = 10
$R_{in} = 22k$

(a)

Typically, $R_1 = 220k$
$R_2 = 22k$
$R_3 = 47k$
$R_4 = 47k$
$C_1 = C_2 = 10\mu F$
Gain = 10
$R_{in} = 22k$

(b)

*Figure 4.3. Inverting amplifier configuration. (a) uses balanced power supplies. Ideally, $R_3$ should equal $R_2$, though differing values are often used. The gain is set by the ratio $R_1/R_2$, and the input resistance is equal to $R_2$. Using an unbalanced power supply (b), the + input is biased to half the supply voltage (15 V in this example) by using equal values for $R_3$ and $R_4$. The gain is again given by $R_1/R_2$. Coupling capacitors are needed because of the d.c. bias conditions*

Circuits for non-inverting amplifiers are shown in *Figure 4.4*. Non-inverting amplifiers also make use of negative feedback to stabilise the working conditions in the same way as the inverting amplifier circuits, but the signal input is now to the (+) input terminal. The gain is

$$G = \frac{R_1 + R_2}{R_2}$$

and the circuit is sometimes referred to as the 'voltage-follower with gain'. The input resistance is high, usually around 1 MΩ, for the dual-supply version, though the bias resistors (*Figure 4.4b*) reduce this to a few hundred kΩ.

Typical values:
$R_1$ = 220k
$R_2$ = 18k
$G$ $= \dfrac{238}{18} = 13\cdot 2$
$R_{in}$ = 1M approx.

(a)

Typical values:
$R_1$ = 220k
$R_2$ = 18k
$R_3$ = 220k
$R_4$ = 220k
$C_1 = C_2 = 1\,\mu F$
$G = 13\cdot 2$
$R_{in}$ = 100k($R_3$ in parallel with $R_4$) approx.

(b)

Figure 4.4. *Non-inverting amplifiers. Using a balanced power supply (a), only two resistors are needed, and the voltage gain is given by* $\dfrac{R_1 + R_2}{R_2}$. *The input resistance is very high. When an unbalanced supply (b) is used, a capacitor $C_2$ must be connected between $R_2$ and earth to ensure correct feedback of signal without disturbing bias. The input resistance is now lower because of $R_3$ and $R_4$ which as far as signal voltage is concerned are in parallel*

*Figure 4.5* shows the 741 used as a differential amplifier, though with a single-ended output. The gain is set by the ratio $R_1/R_2$ as before — note the use of identical resistors in the input circuits to preserve balance.

$$\frac{R_1}{R_2} = \frac{R_4}{R_3}$$

Typically:
$R_1 = R_4 = 220k$
$R_2 = R_3 = 22k$
$G = 10$
$R_{in} = 22k$

*Figure 4.5. Differential amplifier application. Both inputs are used for signals which must be in antiphase (balanced about earth). Any common-mode signals (in phase at both inputs) are greatly attenuated*

## General notes on op-amp circuits

The formulae for voltage gain hold for values of gain up to several hundred times, because the gain of the op-amp used *open-loop* (without feedback) is very high, of the order of 100 000 (100 dB). The maximum load current is about 10 mA, and the maximum power dissipation 400 mW. The 741 circuit is protected against damage from short-circuits at the output, and the protection circuits will operate for as long as the short-circuit is maintained.

The frequency range of an op-amp depends on two factors, the gain-bandwidth product for small signals, and the slew rate for large signals. The gain-bandwidth product is the quantity, $G \times B$, with $G$ equal to voltage gain (not in dB) and $B$ the bandwidth upper limit in Hz. For the 741, the $G-B$ factor is typically 1 MHz, so that, in theory, a bandwidth of 1 MHz can be obtained when the voltage gain is unity, a bandwidth of 100 kHz can be attained at a gain of ten, a bandwidth of 10 kHz at a gain of 100 times, and so on. This trade-off is usable only for small signals, and cannot necessarily be applied to all types of operational amplifiers. Large amplitude signals are further limited by the slew rate of the circuits within the amplifier. The slew rate of an amplifier is the value of maximum rate of change of output voltage

$$\text{slew rate} = \frac{\text{maximum voltage change}}{\text{time needed}}$$

Units are usually volts per microsecond.

Because this rate cannot be exceeded, and feedback has no effect on slew rate, the bandwidth of the op-amp for large signals, sometimes called the power bandwidth, is less than that for small signals. The slew rate limitation cannot be corrected by the use of negative feedback; in fact negative feedback acts to increase distortion when the slew rate limiting action starts, because the effect of the feedback is to increase the rate of change of voltage at the input of the amplifier whenever the rate is limited at the output. This accelerates the overloading of the amplifier, and can change what might be a temporary distortion into a longer-lasting overload condition.

The relationship between the sine wave bandwidth and the slew rate, for many types of operational amplifier is:

$$\text{Max. slew rate} = 2 \pi f_{max} E_{peak}$$

or

$$f_{max} = \frac{\text{max. slew rate}}{2 \pi E_{peak}}$$

where slew rate is in units of volts per second (*not* V/$\mu$s), $f_{max}$ is the maximum full-power frequency in Hz, and $E_{peak}$ is the peak voltage of the output sine wave.

This can be modified to use slew rate figures in the more usual units of V/$\mu$s, with the answer in MHz. For example, a slew rate of 1.5 V/$\mu$s corresponds to a maximum sine wave frequency (at 10 V output) of

$$f_{max} = \frac{1.5}{2 \pi \times 10} \text{ MHz} = 0.023 \text{ MHz or } 23 \text{ kHz.}$$

Slew rate limiting arises because of internal stray capacitances which must be charged and discharged by the current flowing in the transistors inside the i.c.: improvement is obtainable only by redesigning the internal circuitry. The 741 has a slew rate of about 0.5 V/$\mu$s, corresponding to a power bandwidth for 12 V peak sine wave signals of about 6.6 kHz. The slew rate limitation makes op-amps unsuitable for applications which require fast-rising pulses, so that a 741 should not be used as a signal source or feed (interface) with digital circuitry, particularly TTL circuitry, unless a Schmitt trigger stage is also used.

Higher slew rates are obtainable with more modern designs of op-amps; for example, the Fairchild LS201 achieves a slew rate of 10 V/$\mu$s.

### Other operational amplifier circuits

*Figures 4.6* to *4.9* illustrate circuits other than the straightforward voltage amplifier types. *Figure 4.6* shows two versions of follower circuit with no voltage gain, but with useful characteristics. The unity gain inverter will provide an inverted output of exactly the same amplitudes as the input signal, subject to slew rate limitations. The non-inverting circuit, or voltage follower, performs the same action as the familiar emitter-follower, having a very high input resistance and a low output resistance. For this type of circuit, the action of the feedback causes both inputs to change voltage together, as a common-mode signal

Typical values:
$R_1 = R_2 = 330k$
$G = 1$
$R_{in} = 330k$
$R_{out} = 100R$ } approx.

(a)

Typical values:
$R_2 = R_3 = 470k$
$R_1 = 220k$
$G = 1$
$R_{in} = 220k$
$R_{out} = 100R$ } approx.

(b)

*Figure 4.6. The voltage follower. The gain is unity, with high input resistance and low output resistance. The dual-voltage supply version (a) uses only two resistors whose values are not critical. Ideally, $R_1$ should equal $R_2$, and both should be high values so that the input resistance is high, equal to $R_2$. The single supply voltage version (b) uses three resistors, with $R_2 = R_3$ and $R_1$ made equal to $R_2/2$, which is also the value of input resistance*

would, so that any restrictions on the amplitude of common-mode signals (see the manufacturer's sheets) will apply to this circuit. *Figure 4.7* shows two examples of a 741 as it is used in a variety of 'shaping' circuits in which the gain/frequency or gain/amplitude graph is intended to be non-linear.

The use of op-amps for switching circuits is limited by the slew rate, but the circuits shown in *Figures 4.8* and *4.9* are useful if fast-rising or falling waveforms are not needed.

(a)

Wien bridge values R,C determine
frequency

(b)

Figure 4.7. Using the 741 in circuits which are not linear
amplifiers. (a) A constant-output amplifier. Because the
diodes will permit feedback of voltages whose amplitude is
enough to allow the diodes to conduct, the output voltage is
limited to about this amplitude, but without excessive
clipping. The gain is very large for small input signals, and
very small for large input signals. (b) The Wien bridge in the
feedback network causes oscillation. The waveform is a sine
wave only if the gain is carefully controlled by making
$R_1/R_2 = 3$, and this is usually done by making $R_2$ a
thermistor whose resistance value decreases as the voltage
across it increases. The frequency of oscillation is given by

$$f = \frac{1}{2 \pi RC}$$

## Current differencing amplifiers

A variation on the op-amp circuit uses current rather than voltage input signals, and is typified by the National Semiconductor LM3900. In this i.c., which contains four identical op-amps, the + and − inputs are current inputs, whose voltage is generally about +0.6 V when correctly biased. A single-ended power supply is used, and the output voltage can reach to within a fraction of a volt of the supply limits. The output voltage is proportional to the difference between the *currents* at the

*Figure 4.8. A 741 used as an astable multivibrator; a single power supply voltage version is illustrated. The use of $R_2$, $VR_1$, $R_3$ in the positive feedback path sets the + input to a definite fraction of the output voltage above or below the normal half-supply-voltage. When the output changes over, $C_1$ charges through $R_1$ until the − input reaches the same voltage as the + input upon which the circuit switches over. The voltage change at the + input is then rapid, but the voltage at the − input cannot change until $C_1$ has charged again. The frequency is therefore set by the time constant $C_1R_1$ and also by the setting of $VR_1$*

two inputs, so that bias conditions are set by large value resistors. *Figure 4.10* shows a typical amplifier circuit, in which the current into the (+) input is set by $R_1$, whose value is 2.2 MΩ. Because the ideal bias voltage for the output is half of supply voltage, a 1 MΩ resistor is used connected between the output and the (−) input. In this way, the currents to the two inputs are identical, and the amplifier is correctly biased. The advantages of this type of op-amp are now being recognised, and an equivalent to the LM3900 is now available from Motorola.

## Other linear amplifier i.c.s

A very large variety of i.c.s intended for a.f., i.f. and r.f. amplifiers can be obtained. For any design work, the full manufacturer's data sheet pack must be consulted, but a few general notes can be given here. A.F. i.c. circuits use direct coupling internally, because of the difficulty of fabricating capacitors of large value on to silicon chips, but the high gains which are typical of operational amplifiers are not necessary for most a.f. applications. Faster slew rates and greater open-loop bandwidths can therefore be attained than is practicable using op-amps.

*Figure 4.9. A 741 monostable. With no input, the output voltage is high, which causes the + input voltage to be higher than the voltage, $V_s/2$ set by $R_5$ and $R_6$ (equal values). Because of $D_1$, the − input voltage cannot rise to the same value as the + input. A negative pulse at the + input causes the output voltage to drop rapidly, taking the + input voltage low. The − input voltage then drops at a rate determined by the time constant $C_1 R_1$. When the − input voltage equals the + input voltage, the circuit switches back, and the diode $D_1$ conducts to 'catch' the − input voltage and so prevent continuous oscillation*

Many a.f. i.c.s use separate chips for preamplifier and for power amplifier uses, with separate feedback loops for each. Frequency correcting networks composed of resistors and capacitors are usually needed to avoid oscillation, and heatsinks will be needed for the larger power amplifier i.c.s. The need for external volume, stereo balance, bass and treble controls, along with feedback networks, makes the circuitry rather more involved than some other i.c. applications.

*Figure 4.11* shows two a.f. circuits examples. Note that the stability of these audio i.c.s is often critical, and decoupling capacitors, as specified by the manufacturers, must be connected as close to the i.c.

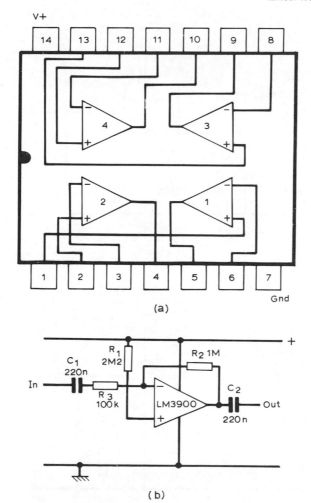

Figure 4.10. The current differencing amplifier, or Norton
Op-amp. (a) Pinout for the LM3900, which contains four
amplifiers in a single fourteen-pin package. (b) Typical
amplifier circuit. Note the high resistor values

pins as possible. For stability reasons, also, stripboard construction is
extremely difficult with some i.c. types, and suitable printed-circuit
boards should be used.

I.F. and r.f. amplifier circuits contain untuned wideband amplifier
circuits to which tuning networks, which may be *LC* circuits or trans-
filters, may be added. It is possible to incorporate r.f., mixer, i.f. and

(a)

(b)

Figure 4.11. Audio amplifier i.c.s. (a) The MC1303 preamplifier is a dual unit for stereo use — the pin numbers in brackets are for the second section. Inputs up to 5 mV can be accepted, and the circuit here is shown equalised for a magnetic pickup. The output is 250 mV with a 5 mV input at a distortion level of about 0.1%. (b) The LM383 power amplifier uses a five-pin TO 220 package. The power output is 7 W into 4 ohms, with a distortion level of 0.2% at 4 W output. The maximum power dissipation is 15 W when a 4 °C/W heatsink is used demodulator stages into a single i.c., but generally only when low-frequency r.f. and i.f. are used. A very common scheme for f.m. radio receivers is to use a discrete component tuner along with i.c. i.f. and demodulator stages, using the usual 10.7 MHz i.f. *Figure 4.12* shows an example of such an i.f. stage. Once again, when a large amount of gain

*Figure 4.12. An i.f. detector i.c. for use in 10.7 MHz stereo f.m. i.f. stages. The minimum input for limiting is 100 µV, and the volume control range (operating on d.c.) is 80 dB. The audio output is 1.4 V r.m.s. with a signal of 15 kHz deviation*

is attained in one i.c. stability is a major problem, and the manufacturer's advice on decoupling must be carefully followed. At the higher frequencies, the physical layout of components is particularly important, so that p.c.b.s intended for the TBA750 i.c. (and similar) should be used rather than stripboards.

## Phase locked loops

The phase-locked loop is a type of linear i.c. which is now being used to a considerable extent. The block diagram of the circuit is outlined in

*Figure 4.13* and consists of a voltage-controlled oscillator, a phase-sensitive detector, and comparator units. The oscillator is controlled by external components, so that the frequency of oscillation can be set by a suitable choice of these added components. An input signal to the Phase Locked Loop (PLL) is compared in the phase-sensitive detector to the frequency generated in the internal oscillator, and a voltage output obtained from the phase-sensitive detector. Provided that the input frequency is not too different from the internally generated

*Figure 4.13. The phase-locked loop (p.l.l.) block diagram. The pin numbering is for the LM565. The signal input can be to pin 2 or 3 in this i.c., and in normal use pins 4 and 5 are linked*

frequency (within the 'pull-in' range), the voltage from the phase-sensitive detector can then be used to correct the oscillator frequency until the two signals are at the same frequency and in the same phase. Either the oscillator or the correcting voltage may be used as an output. The circuit can be used, for example, to remove any traces of amplitude modulation from an input signal, since the output (from the internal oscillator) is not affected by the amplitude of the input signal, but is locked to its frequency and phase. The circuit may also be used as an f.m. demodulator, since the control voltage will follow the modulation of an f.m. input in its efforts to keep the oscillator locked in phase.

(b)

*Figure 4.14. P.L.L. circuits. (a) Oscillator with fundamental and second harmonic outputs. (b) F.M. demodulator — the component values must be calculated with reference to the i.f. frequency which is used — this cannot be as high as the normal 10.7 MHz because the operating frequency limit of this i.c. is 500 kHz*

## Voltage stabiliser i.c.s

The ease with which zener diode junctions and balanced amplifiers may be constructed in integrated form, together with the increasing demand for stabilised supplies and the steady increase in the power which can be dissipated from i.c.s due to improved heat-sinking methods, has led to the extensive use of i.c. voltage stabiliser circuits, to the extent that discrete component stabilisers are almost extinct. The decisive factor is that i.c.s can include such features as overload and short-circuit protection at virtually no extra cost, using an elaborate circuit such as that of the SGS-ATES TBA625A shown in *Figure 4.15*. The overload

Figure 4.15. Voltage regulation. (a) Use of an i.c. regulator, the TBA625A, is illustrated. (b) Foldback overcurrent protection — at excessive currents the voltage output and current output both drop as indicated in the graph

protection is usually of the 'foldback' type, illustrated by the characteristic of *Figure 4.15b*, in which excessive current causes the output voltage to drop to zero with a much smaller current flowing. *Figure 4.16* shows circuit applications — note that a fixed voltage regulator can be used to provide an adjustable output, and higher current operation can be achieved by adding power transistors to the circuit. Voltage stabiliser i.c.s are available for all the commonly used voltage levels.

Motor-speed controllers are a more specialised form of stabiliser circuit, and are used to regulate the speed of d.c. motors in record players, tape and cassette recorders, and model motors. A typical application is shown in *Figure 4.17*.

One very important class of linear i.c.s is concerned with television circuitry. The development of linear i.c.s has been such that virtually every part of a TV circuit with the exceptions of the tuner head and the horizontal output stage can now be obtained in i.c. form. Because of the specialised nature of such circuits, the reader is referred to the

Figure 4.16. Extending the range of fixed-voltage regulators. (a) Extending voltage range, (b) extending current range

Figure 4.17. A motor-speed control i.c. The values of $VR_1$ and $R_1$ must be calculated with reference to the resistance of the motor windings

*Figure 4.18. The TDA 1270 used for vertical deflection in a 12 in portable TV receiver*

manufacturer's handbooks for further information, but as an example of the uses of such i.c.s, *Figure 4.18* shows a TDA1270 (SGS-ATES) which performs the functions of vertical oscillator, timebase generator and output stage.

### The 555 timer

This circuit is generally classed among linear circuits because it uses op-amp circuits as comparators. The purpose of the timer is to generate time delays or waveforms which are very well stabilised against voltage changes. A block diagram of the internal circuits is shown in *Figure 4.19*.

A negative-going pulse at the trigger input, pin 2, makes the output of comparator (2) go high. The internal resistor chain holds the (+) input of comparator (2) at one third of the supply voltage, and the (−) input of comparator (1) at two-thirds of supply voltage, unless pin 5 is connected to some different voltage level. The changeover of comparator (2) causes the flip-flop to cut off $Tr_1$, and also switch the output stage to its high-voltage output state. With $Tr_1$ cut off, the external capacitor $C$ can charge through $R$ (also external) until the voltage at pin

6 is high enough, equal to 2/3 of the supply voltage, to operate comparator (1). This resets the flip-flop, allows $Tr_1$ to conduct again, so discharging $C$, and restores the output to its low voltage state. Resetting is possible during the timing period by applying a negative pulse to the reset pin, number 4.

Figure 4.19. *The 555 timer block diagram. R\* and C are external components which are added in most applications of the timer*

The triggering is very sensitive, and some care has to be taken to avoid unwanted triggering pulses, particularly when inductive loads are driven. Retriggering caused by the back-e.m.f. pulse from an inductive load is termed 'latch-up', and can be prevented by the diode circuitry shown in *Figure 4.20*.

*Figure 4.20. A relay-timer circuit using the 555. On pressing the switch, the relay is activated for a time determined by the setting of the 470 kΩ variable and the capacitor value selected by the switch. Note the use of diodes to prevent latch-up and damage to the i.c. when the relay is switched off*

*Figure 4.21. An astable pulse generator, with variable frequency output controlled by the 100 kΩ potentiometer. The capacitor C can be a switched value if desired*

Two typical circuits are shown in *Figures 4.21* and *4.22*. The timer is available from several manufacturers, all using the same 555 number though prefixed with different letter combinations which indicate the manufacturer.

# Chapter 5

# Digital I.C.s

## Basic logic notes

Unlike linear circuits, digital circuits process signals which consist of only two significant voltage levels, labelled logic 0 and logic 1. Most logic systems use positive logic, in which logic 0 is represented by zero volts, or a low voltage, below 0.5 V; and logic 1 is represented by a higher voltage. Changes of level from 0 to 1 or 1 to 0 are made as quickly as possible, since slow changes can cause faulty operation of some types of logic circuits.

The use of two logic levels naturally leads to the use of a scale-of-two or binary scale for counting. In a binary scale, the only digits used are 1 and 0, with the placing of the 1 indicating what power of 2 is represented. *Table 5.1* shows methods of converting to and from binary numbers and decimal numbers.

In addition, because large binary numbers are awkward to handle, and difficult to copy without error, hexadecimal (or hex) numbers are used for many applications, particularly in microprocessor machine code (see later). Hex coding is used when binary numbers occur in groups of four, eight (called a byte) or multiples, and the conversions are shown in *Table 5.2*. The use of hex coding makes the tabulation of binary numbers considerably simpler, but the circuits to which the hex codes refer will still operate in binary.

Note in this context that a three-state or tri-state device does not imply a third logic voltage. This description refers to a circuit which can be switched to a high impedance at an input or output (or both) so that it does not affect or respond to the voltages of signals connected to it.

## Table 5.1. CODE CONVERSION

### DECIMAL-TO-BINARY CONVERSION

Write down the decimal number. Divide by two, and write the result underneath, with the remainder, 0 or 1, at the side. Now divide by two again, placing the new (whole number) result underneath, and the remainder, 0 or 1, at the side. Repeat until the last figure (which will be 2 or 1) has been divided, leaving zero.

Now read the remainders in order from the foot of the column upwards to give the binary number.

**Example:**  Convert decimal 1065 into binary:

| | |
|------|---|
| 1065 | |
| 532 | 1 |
| 266 | 0 |
| 133 | 0 |
| 66 | 1 |
| 33 | 0 |
| 16 | 1 |
| 8 | 0 |
| 4 | 0 |
| 2 | 0 |
| 1 | 0 |
| 0 | 1 |

Binary number is 10000101001

### BINARY-TO-DECIMAL CONVERSION

Using the table of binary powers, which shows the values of successive powers of two, starting from the right-hand side of the binary number (the least significant figure). Write down the decimal equivalent for each 1 in the binary number, and then add.

**Example:**  1  0  0  0  0  0  0  1  0  0  1

| | Decimal |
|---|---|
| | 1 |
| | 8 |
| | 32 |
| | 1024 |
| Total | 1065 |

| Power of two | Decimal | Power of two | Decimal |
|---|---|---|---|
| 0 | 1 | 10 | 1 024 |
| 1 | 2 | 11 | 2 048 |
| 2 | 4 | 12 | 4 096 |
| 3 | 8 | 13 | 8 192 |
| 4 | 16 | 14 | 16 384 |
| 5 | 32 | 15 | 32 768 |
| 6 | 64 | 16 | 65 536 |
| 7 | 128 | 17 | 131 072 |
| 8 | 256 | 18 | 262 144 |
| 9 | 512 | 19 | 524 288 |

## Table 5.2. BINARY, HEXADECIMAL AND DECIMAL

| 4-bit Binary | Decimal | Hexadecimal |
|---|---|---|
| 0000 | 0 | 0 |
| 0001 | 1 | 1 |
| 0010 | 2 | 2 |
| 0011 | 3 | 3 |
| 0100 | 4 | 4 |
| 0101 | 5 | 5 |
| 0100 | 6 | 6 |
| 0111 | 7 | 7 |
| 1000 | 8 | 8 |
| 1001 | 9 | 9 |
| 1010 | 10 | A |
| 1011 | 11 | B |
| 1100 | 12 | C |
| 1101 | 13 | D |
| 1110 | 14 | E |
| 1111 | 15 | F |

To convert binary to hex: arrange the binary numbers in four-bit groups starting from the right-hand side (the least significant digit). Convert each group to hex, using the table above.

**Examples:** 01011101 converts to 5D,   11100011 converts to E3

To convert hex to binary: write down the equivalent for each number, using the table above.

**Examples:** E1 = 11100001,   6F = 01101111

---

Combinational logic circuits are those in which the output at any time is determined entirely by the combination of input signals which are present at that time. A combinational logic circuit can be made to produce a logic 1 at its output only for some set pattern of 1s and 0s at various inputs, just as a combination lock will open only when the correct pattern of numbers has been set. One of the major uses for

AND        OR        NAND        NOR        NOT

(a)

(b)

*Figure 5.1. Gate symbols. (a) International (U.S. MIL) symbols, used in magazine articles, most books and manufacturer's manuals, (b) British Standard (B.S.) symbols used for TEC and C & G examinations and textbooks* combinational logic circuits is in recognising patterns, whether these be binary numbers or other information in binary form.

Any required combinational logic circuit can be built up from a few basic circuits, the choice of which is dictated more by practical than by theoretical considerations. The three most basic circuits are those of the

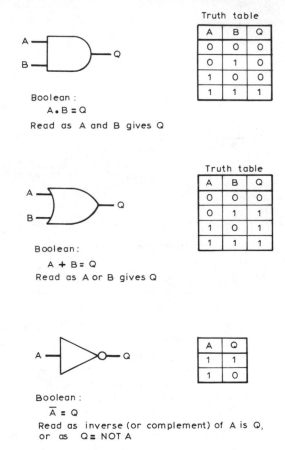

Boolean :

A.B = Q

Read as A and B gives Q

Truth table

| A | B | Q |
|---|---|---|
| 0 | 0 | 0 |
| 0 | 1 | 0 |
| 1 | 0 | 0 |
| 1 | 1 | 1 |

Boolean:

A + B = Q

Read as A or B gives Q

Truth table

| A | B | Q |
|---|---|---|
| 0 | 0 | 0 |
| 0 | 1 | 1 |
| 1 | 0 | 1 |
| 1 | 1 | 1 |

Boolean :

$\overline{A}$ = Q

Read as inverse (or complement) of A is Q,
or as   Q = NOT A

| A | Q |
|---|---|
| 1 | 1 |
| 1 | 0 |

*Figure 5.2. Truth tables and Boolean expressions for three basic gates*

AND, OR and NOT gates; the NOT gate is also referred to as an inverter or complementer. These are represented in circuits by the symbols shown in *Figure 5.1*; the international symbols are much more commonly used. The internal circuitry of the i.c.s is not usually shown, since the action of the circuits is standardised. All that need be known about the internal circuits is the correct level of power supplies, driving signals and output signals, together with any handling precautions.

The action of a logic gate or circuit can be described by a truth table or by a Boolean expression. A truth table shows each possible input to the logic circuit with the output which such a set of inputs produces; examples are shown in *Figure 5.2*. A Boolean expression is a much shorter way of showing the action, using the symbols + to mean OR and . to mean AND. The action of a four-input AND gate, for example, can be written as $A.B.C.D = 1$; a truth table for this gate would take up half a page, because the number of lines of truth table is equal to $2^n$, where $n$ is the number of inputs.

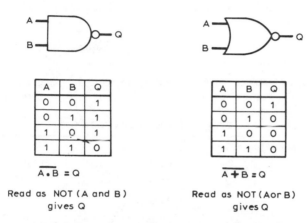

| A | B | Q |
|---|---|---|
| 0 | 0 | 1 |
| 0 | 1 | 1 |
| 1 | 0 | 1 |
| 1 | 1 | 0 |

$$\overline{A.B} = Q$$

Read as NOT (A and B)
gives Q

| A | B | Q |
|---|---|---|
| 0 | 0 | 1 |
| 0 | 1 | 0 |
| 1 | 0 | 0 |
| 1 | 1 | 0 |

$$\overline{A+B} = Q$$

Read as NOT (A or B)
gives Q

*Figure 5.3. NAND and NOR gate truth tables and Boolean expressions*

The NAND and NOR gates shown in *Figure 5.3* combine the actions of the AND and OR gates with that of an inverter. These gates are simpler to produce, and either can be used as an inverter so that for practical purposes these are more common than the ordinary AND and OR gates. Another type of circuit which is used to such an extent that it is available in i.c. form is the exclusive-OR (X-OR) gate. The truth table for a two-input X-OR gate is shown in *Figure 5.4*, in comparison

| A | B | Q |
|---|---|---|
| 0 | 0 | 0 |
| 0 | 1 | 1 |
| 1 | 0 | 1 |
| 1 | 1 | 0 |

Boolean: $(\overline{A.B}).(A+B) = Q$

*Figure 5.4. Exclusive-OR (X-OR) gate. Symbols, truth table and Boolean expression*

with that for a simple OR-gate. The Boolean expression is $Q = \overline{(A + B)}$. $(A.B)$, where the bar above the brackets indicates NOT (= inverse). The expression $A.B$ will be 1 when $A = 1$ and $B = 1$, and the inverse bar indicates that the output must be zero for this state. This quantity is ANDED with (A or B), so as to make the output zero for $A = 1$ and $B = 1$.

The advantages of setting out logic circuit requirements as Boolean expressions rather than in truth tables are:

(1)    The Boolean expressions are considerably quicker to write.

(2)    The Boolean expressions can be simplified (often) by applying a set of rules.

(3)    The Boolean expressions can usually indicate what combination of single gates will be needed for the circuit.

The rules of Boolean algebra (invented, incidentally, long before digital logic circuits existed) are shown in *Table 5.3*. The usefulness of these

**Table 5.3. BOOLEAN ALGEBRA**

1. Definitions of gate action

| | |
|---|---|
| $0 + 0 = 0$ | $0.0 = 0$ |
| $0 + 1 = 1$ | $0.1 = 0$ |
| $1 + 0 = 1$ | $1.0 = 0$ |
| $1 + 1 = 1$ | $1.1 = 1$ |

2. For gates to which one input is a variable A (which can be 0 or 1)

| | |
|---|---|
| $A + 0 = A$ | $A.0 = 0$ |
| $A + 1 = 1$ | $A.1 = A$ |
| $A + A = A$ | $A.A = A$ |
| $A + \overline{A} = 1$ | $A.\overline{A} = 0$ |
| $\overline{\overline{A}} = A$ | |

3. For more than one variable (each of which can be 0 or 1)

| | |
|---|---|
| $A + B = B + A$ | $A.B = B.A$ |
| $(A + B) + C = A + (B + C)$ | $(A.B).C = A.(B.C)$ |
| $(A.B) + (A.C) = A.(B + C)$ | $(A + B).(A + C) = A + (B.C)$ |

4. De Morgan's theorem

| | |
|---|---|
| $\overline{A.B} = \overline{A} + \overline{B}$ | $\overline{A + B} = \overline{A}.\overline{B}$ |

rules is that they may be applied to simplify an expression, so saving considerably in the number of logic gates that are needed to carry out a logic operation. For example, the proposed gate system in *Figure 5.5* carries out the process which in Boolean algebra becomes

$$(A . B . C) + A . (\overline{B} + \overline{C})$$

Leaving the first term unchanged, we can apply Rule 4 to the second term

$(A . B . C) + A . (\overline{B . C})$

Now taking out the common factor A changes the expression to

$A . (B . C + \overline{B . C})$      (Rule 3)

$= A . 1$             (Rule 2)

$= A$               (Rule 2)

From this, it appears that only the A signal is needed. An experienced designer might, in fact, be able to deduce that the B and C signals were

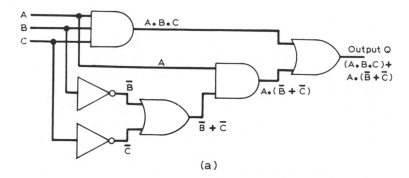

(a)

| A | B | C | A.B.C | $\overline{B}$ | $\overline{C}$ | $\overline{B}+\overline{C}$ | $A.(\overline{B}+\overline{C})$ | Q |
|---|---|---|-------|---|---|------|----------|---|
| 0 | 0 | 0 | 0 | 1 | 1 | 1 | 0 | 0 |
| 0 | 0 | 1 | 0 | 1 | 0 | 1 | 0 | 0 |
| 0 | 1 | 0 | 0 | 0 | 1 | 1 | 0 | 0 |
| 0 | 1 | 1 | 0 | 0 | 0 | 0 | 0 | 0 |
| 1 | 0 | 0 | 0 | 1 | 1 | 1 | 1 | 1 |
| 1 | 0 | 1 | 0 | 1 | 0 | 1 | 1 | 1 |
| 1 | 1 | 0 | 0 | 0 | 1 | 1 | 1 | 1 |
| 1 | 1 | 1 | 1 | 0 | 0 | 0 | 0 | 1 |

(b)

*Figure 5.5. Analysing the action of a gate system. (a) The text shows the method using Boolean algebra, the truth table shows that the Q output column corresponds exactly with the A input column. The truth table method is simple, but very tedious, because the number of lines of truth table are equal to $2^n$, where n is the number of inputs (A, B, C, etc.)*

redundant by an inspection of the gate circuit, but the Boolean algebra is often faster to write, if nothing else!

By way of another example, consider the gates of *Figure 5.6*, which implement the Boolean expression

$$Q = \overline{A} \cdot C \cdot + \overline{A} \cdot B \cdot D + A \cdot C \cdot D + A \cdot B \cdot D$$

Taking out the common factor D, this becomes

$$Q = D \cdot (\overline{A} \cdot C + \overline{A} \cdot B + A \cdot C + A \cdot B)$$

and taking out the next common factor $(C + B)$, this transforms to

$$Q = D \cdot (\overline{A} \cdot (C + B) + A \cdot (C + B))$$

which is

$$Q = D \cdot (C + B) \cdot (\overline{A} + A)$$
$$= D \cdot (C + B), \text{ since } \overline{A} \text{ or } A \text{ must be logic 1.}$$

(a)

| A | B | C | D | $\overline{A}$ | $\overline{A} \cdot C \cdot D$ | $\overline{A} \cdot B \cdot D$ | $A \cdot C \cdot D$ | $A \cdot B \cdot D$ | Q |
|---|---|---|---|---|---|---|---|---|---|
| 0 | 0 | 0 | 0 | 1 | 0 | 0 | 0 | 0 | 0 |
| 0 | 0 | 0 | 1 | 1 | 0 | 0 | 0 | 0 | 0 |
| 0 | 0 | 1 | 0 | 1 | 0 | 0 | 0 | 0 | 0 |
| 0 | 0 | 1 | 1 | 1 | 1 | 0 | 0 | 0 | 1 |
| 0 | 1 | 0 | 0 | 1 | 0 | 0 | 0 | 0 | 0 |
| 0 | 1 | 0 | 1 | 1 | 0 | 1 | 0 | 0 | 1 |
| 0 | 1 | 1 | 0 | 1 | 0 | 0 | 0 | 0 | 0 |
| 0 | 1 | 1 | 1 | 1 | 1 | 1 | 0 | 0 | 1 |
| 1 | 0 | 0 | 0 | 0 | 0 | 0 | 0 | 0 | 0 |
| 1 | 0 | 0 | 1 | 0 | 0 | 0 | 0 | 0 | 0 |
| 1 | 0 | 1 | 0 | 0 | 0 | 0 | 0 | 0 | 0 |
| 1 | 0 | 1 | 1 | 0 | 0 | 0 | 1 | 0 | 1 |
| 1 | 1 | 0 | 0 | 0 | 0 | 0 | 0 | 0 | 0 |
| 1 | 1 | 0 | 1 | 0 | 0 | 0 | 0 | 1 | 1 |
| 1 | 1 | 1 | 0 | 0 | 0 | 0 | 0 | 0 | 0 |
| 1 | 1 | 1 | 1 | 0 | 0 | 0 | 1 | 1 | 1 |

(b)

*Figure 5.6. A gate system (a) whose action is analysed by Boolean algebra in the text. The truth table (b) shows that the output is logic 1 when D is 1 and C or B also 1*

For really large and complex circuits, Boolean algebra can be tedious and difficult, but other approaches are equally tedious and difficult — there is no easy way of dealing with complex circuits.

Note that in circuits which use an elaborate sequence of gates, problems can arise because of the small but significant time delay which

*Figure 5.7. A circuit for suppressing unwanted short pulses. This also inverts, so that a second Schmitt inverter may be needed after the Schmitt gate*

is introduced by each gate. These 'race hazard' problems arise when the final gate of a set is switched by signals which have passed through different numbers of gates in their different paths. A time delay between these paths can cause a brief unwanted pulse, of duration equal to the time delay between the input paths, at the output. This unwanted pulse may cause problems, particularly in a counting circuit which follows the gate circuits. Methods of eliminating race hazards are too complex to discuss in this book, but *Figure 5.7* shows a simple unwanted-pulse suppressor which can be used for comparatively slow logic systems. The output is integrated by *R, C*, using a time constant chosen so that a short pulse (50 ns or so) will have little effect, but the delay which the circuit causes to a longer pulse is not serious. The sharp rise and fall times of the logic signals then have to be restored by using a Schmitt NAND gate, IC1.

### Sequential logic

Sequential logic circuits change output when the correct sequence of signals appears at the inputs. The simplest of the sequential logic circuits is the R-S (or S-R) flip-flop, or latch, circuit shown in symbol form in *Figure 5.8a* with one possible circuit using NAND gates. The truth table for this device is shown in *Figure 5.8b*, from which it can be seen that it is the sequence of signals at the two inputs rather than the signal levels themselves which decide the output. Note that the outputs Q and $\overline{Q}$ are intended to be complementary (one must be the inverse

**Truth table**

| R | S | Q | Q̄ |
|---|---|---|---|
| 0 | 0 | 1✶ | 1✶ |
| 0 | 1 | 1 | 0 |
| 1 | 1 | 1 | 0 |
| 1 | 0 | 0 | 1 |
| 1 | 1 | 0 | 1 |

✶Undesirable state

The R =0, S=0 state must be avoided because we normally want Q̄ to be the inverse of Q

(b)

*Figure 5.8. The R-S flip-flop. (a) Symbol and one possible circuit (NOR gates can also be used). (b) Truth table. The output for R=1, S=1 depends on the previous values of R and S*

of the other), so that the state R = 0, S = 0 must never be allowed, since it results in Q = 1, Q̄ = 1.

The R-S flip-flop is used to lock or latch information, one bit to each R-S latch, because the output of the latch is held fixed when both

(a)

(b)

*Figure 5.9. A 'switch-debounce' circuit using an R-S flip-flop (a). An alternative arrangement which needs only an on/off switch type is shown in (b) — this uses a Schmitt inverter rather than the R-S flip-flop* inputs are at logic 1 in the state which existed just before the second input went high. One typical application is to suppress the effects of contact bounce in mechanical switches, as shown in *Figure 5.9*. The effect of bounce in this arrangement is to raise both inputs of the latch to logic 1, leaving the outputs unaffected.

The applications of the simple R-S latch are rather limited, and most sequential logic circuits make use of the principle of clocking. A clocked circuit has a terminal, the clock (CK) input, to which

| D | $Q_n$ | $Q_n+1$ |
|---|---|---|
| O | O | O |
| O | 1 | O |
| 1 | O | 1 |
| 1 | 1 | 1 |

*Figure 5.10. Symbol for D-type flip-flop and truth table. In the truth table, $Q_n$ denotes the output before the clock pulse, and $Q_n+1$ the output after the clock pulse*

rectangular pulses, the clock pulses) can be applied. The circuit actions, which is determined by the other inputs, takes place only in the time of the clock pulse, and may be synchronised to the leading edge, the trailing edge, or the time when the clock pulse is at logic 1, according to the design of the circuit.

The D-type flip-flop, shown in *Figure 5.10*, with its truth table, is one form of D-type clocked flip-flop. The output at Q will become identical to the input at D (for DATA) only at the clock pulse, generally at the leading edge of the pulse, so earning the circuit the name 'edge-triggered'. By connecting the $\overline{Q}$ output back to the D-input, the flip-flop will act as a bistable counter, or divide-by-two circuit, because the voltage at Q will cause a change of state only when the leading edge of the clock-pulse occurs. The $\overline{Q}$ voltage itself will change only a little time after the leading edge of the clock pulse, so that there is no effect on the flip-flop. A few D-type flip-flops change state at the trailing edge of the clock pulse, and some allow the D-input to affect the Q-output for as long as the clock pulse is at logic 1. This latter type is sometimes called a 'transparent latch'.

The J-K flip-flop (*Figure 5.11a*) is a much more flexible design which uses a clock pulse along with two control inputs labelled J and K. The internal circuit makes use of two sets of flip-flops, designated master and slave respectively. At the leading edge of the clock pulse, the master flip-flop changes state under the control of the inputs at J and K. After this time, changes at J and K have no effect on the master flip-flop. There is no change at the output, however, until the trailing

| J | K | $Q_n$ | $Q_n+1$ |
|---|---|---|---|
| 0 | 0 | 0 | 0 |
|   |   | 1 | 1 |
| 0 | 1 | 0 | 0 |
|   |   | 1 | 0 |
| 1 | 0 | 0 | 1 |
|   |   | 1 | 1 |
| 1 | 1 | 0 | 1 |
|   |   | 1 | 0 |

(a)  (b)

| J | K | $Q_n$ | $Q_n+1$ |
|---|---|---|---|
| 0 | 0 | X | no change |
| 0 | 1 | X | 0 |
| 1 | 0 | X | 1 |
| 1 | 1 | X | $\overline{X}$ |

(c)

*Figure 5.11. The J-K flip-flop (a) and its truth table (b). The truth table can be shortened (c) by making use of X to mean 'don't care' — either 0 or 1*

edge of the clock. The use of this system ensures a controllable time delay between the output and the input, which permits signals to be fed back from one flip-flop to another without the risk of instability, since an output signal fed back to a J or K input cannot have any effect until the clock leading edge, and will not itself change again until the clock trailing edge. This avoids the difficulty of having an input changing while the output which it controls also changes.

The remaining two signal inputs of the J-K flip-flop operate independently of the clock. The SET (or PRESET) terminal sets the Q output to logic 1, regardless of other signals, and the RESET or CLEAR terminal sets the output to logic 0, also regardless of other signals. The system must be arranged so that these two cannot operate together.

*Figure 5.12* shows a type of binary counter variously named serial, ripple, or asynchronous. Using four stages of J-K flip-flops, the counter Q outputs will reach 1111 (decimal 15) before resetting. By using gates to detect 1010 (decimal ten), the counter can be forced to reset on the tenth input pulse, so that a decimal count is obtained. Such a set of gated flip-flops forms the basis of a BCD (Binary-Coded Decimal) counter such as the 7490.

*Figure 5.12. A ripple or asynchronous counter using J-K flip-flops. Note that the flip-flops (and the Q outputs) are labelled 0, 1, 2, 3 rather than 1, 2, 3, 4. This is because the first stage counts $2^0$, the second stage $2^1$, the third stage $2^2$ and so on*

*Figure 5.13. A synchronous counter. The input (clock) pulses are taken to each clock terminal, and the changeover of each flip-flop is controlled by the settings of the J, K terminals, which are gated*

*Figure 5.13* shows a synchronous counter, in which each flip-flop is clocked at the same rate, and counting of the clock pulses is achieved by using gated connections to the J and K inputs. The design of such synchronous counters for numbers which are not powers of two is complex and lengthy. The use of a technique called Karnaugh mapping simplifies the procedure to some extent for small count numbers, but is beyond the scope of this book. In any case, either ripple or synchronous counters can now be obtained as complete integrated circuits for most useful count values, making the design procedure unnecessary, and for complex applications, the use of a microprocessor is probably a simpler solution.

*Figure 5.14. Shift register. At each clock pulse, the binary digit at the Q output of each flip-flop is 'shifted' to the next flip-flop in line. The arrangement shown here gives right-shift; left-shift can be arranged by connecting $\overline{Q}$ to J and Q to K between flip-flops. For details and circuits, see Beginner's Guide to Digital Electronics*

All counter circuits can be arranged so as to count in either direction up or down, and complete counters in i.c. form can be obtained which will change counting direction by altering a control voltage to 0 or 1. An example of such a counter is the 74192.

Shift registers, which can be formed from J-K flip-flops connected as shown in *Figure 5.14*, are also obtained in i.c. form. The action of a

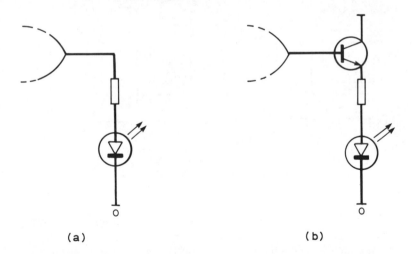

(a)                                    (b)

*Figure 5.15. LED displays. TTL i.c.s can drive the LED directly (a) using a limiting resistor. CMOS i.c.s generally have low current outputs, and a transistor 'interface' circuit (b) is needed. Multiple transistors for this purpose can be obtained in i.c. form, such as the RCA CA8083 (5 NPN) and CA3082 (7 NPN transistors with a common collector connection)* shift register is to pass a logic signal (1 or 0) from one flip-flop to the next in line at each clock pulse. The input signals can be serial, so that one bit is shifted in at each clock pulse, or parallel, switched into each flip-flop at the same time. The output can similarly be serial, taken

(a)   (b)

| Count 0 | Segments illuminated a b c d e f |
|---------|----------------------------------|
| 1 | b c |
| 2 | a b g e d |
| 3 | a b g c d |
| 4 | f g b c |
| 5 | a f g c d |
| 6 | a f e d c g |
| 7 | a b c |
| 8 | a b c d e f g |
| 9 | a b g f c |
| A | a b c f e |
| B | f e g c d |
| C | a f e d |
| D | g e d c b |
| E | a f g e d |
| F | a f g e |

(c)

Figure 5.16. The seven-segment display (a) showing the segment lettering. An eighth segment, the decimal point, is often added, and may be on the right or left side. Hexadecimal letters as well as figures can be displayed provided that 6 and b are differentiated as shown (b) by the use of the a-segment. The truth table (c) shows the segments illuminated for each number in a hex count. Displays may be common anode or common cathode

from one terminal at each clock pulse, or parallel at each flip-flop output. The shift can be left, right or switched in either direction. Some circuits which make use of shift registers are discussed in detail in *Beginner's Guide to Digital Electronics.*

### Displays and decoders

A binary number can be displayed using any visible on/off device; the most convenient is usually the LED (*Figure 5.15*). For decimal number and letter displays (alphanumeric displays), dot matrix displays are used. These consist of an array of LED dots, usually in a 7 × 5 arrangement, which can be illuminated to display a number or letter. A more common arrangement is the seven-segment display (*Figure 5.16*), consisting of bars which can be illuminated. Seven-segment displays may be common-anode, with a common positive connection and each bar illuminated by connecting its terminal to logic zero. The alternative is the common-cathode type, in which the common cathode terminal is earthed and each bar anode is illuminated by taking its voltage to logic 1. A current-limiting resistor must be wired in series with each bar terminal whichever type of system is used.

*Figure 5.17. A strobed or multiplexed display. The decoder i.c.s are operated in sequence by the enable lines (from a counter), which also select the correct display. Only seven segment connections are needed for all the displays, so that fewer pins are needed. The interface i.c.s must be tri-state so that when one is active, driving the segment lines, the others are not affected*

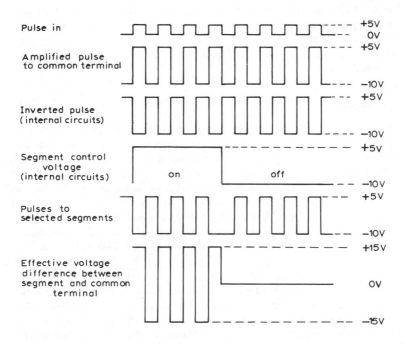

Figure 5.18. An LCD driving circuit and operating pulses. No d.c. must be applied to an LCD display, so that the pulses must be symmetrical about earth. The pulses are applied in phase to segments and common terminal (backplate) when the segments are inactive, but in antiphase to selected segments when a display is wanted

Whether the display is dot-matrix or seven-segment, a decoder i.c. is needed to convert from binary or decimal signals into the correct arrangement of bars or dots. A very common arrangement is to use binary coded decimal (BCD) in which a set of four bits represents each decimal digit, so that BCD-to-seven segment decoders are available; combined counter-decoders are also available. A few hexadecimal decoders exist for seven-segment drive, and Texas make combined decoder-display units. For dot-matrix displays, a read-only memory is usually used in place of a decoder, to take advantage of the much greater range of alphanumeric characters which can be obtained.

In some circuits, to save space or battery current, displays are strobed, meaning that the figures to be displayed are decoded and displayed one after the other, each in its correct place on the display (*Figure 5.17*).

Liquid crystal displays (LCD) are also used in battery-operated equipment. Their advantages include very low power consumption and better visibility under bright lighting, but at the expense of greater circuit complexity, higher cost, and the need for illumination in poor light. The signals from the decoder must be a.c., with no trace of d.c., at a frequency of a few hundred Hz. A typical LCD display circuit is shown in *Figure 5.18*.

## I.C. types

The two main digital i.c. manufacturing methods are described as TTL (Transistor Transistor Logic) and MOS (Metal-Oxide-Silicon) — various types of MOS such as CMOS (complementary p and n channel used), PMOS (positive channel) and NMOS (negative channel) exist. TTL i.c.s use bipolar transistors in integrated form, with input and output circuits which are similar to those shown in *Figure 5.19*. Since the input is always to an emitter, the input resistance is low, and because the base of the input transistor is connected to the +5 V line, the input passes a current of about 1.6 mA when the input voltage is earth, logic 0. If an input is left unconnected, it will 'float' to logic 1, but can be affected by signals coupled by stray capacitance, so that such an input would normally be connected to +5 V through a 1 k$\Omega$ resistor. At the output, a totem-pole type of circuit is used. This can supply a current to a load which is connected between the output and earth (current sourcing), or can absorb a current from a load connected between the output and the +15 V line (current sinking). The normal TTL output stage can source or sink 16 mA, enabling it to drive up to ten TTL inputs. In the language of digital designers, the output stage has a fanout of ten.

TTL i.c.s which use these output stages must never be connected with the outputs of different units in parallel, since with one output stage at logic 1 and another at logic 0, large currents could pass,

*Figure 5.19. TTL circuitry. The example shown is a 2-input NAND gate. The inputs are to emitters of a transistor (in this example, a transistor with two emitters formed on to one base). The output is from a series transistor circuit so that rise and fall times are short*

destroying the output stages. Modified output stages, which have open collector outputs, are available for connecting in parallel; an application which is called a wired OR, since the parallel connections create an OR gate at the output.

Inputs such as SET and RESET (or PRESET and CLEAR) normally have to be taken to logic 0 to operate, this is sometimes indicated by a small circle on the input at the symbol for the device. Clock pulses may affect the circuit either on the leading edge, trailing edge, both edges, or at logic 1; the characteristic sheet for the device must be consulted to make certain which clocking system is in use. The supply

R = 330 Ω to 680Ω

For R = 390Ω

| C | f |
|---|---|
| 100μF | 20Hz |
| 10 μF | 200Hz |
| 1μ F | 2k Hz |
| 100nF | 20kHz |
| 10 nF | 200kHz |
| 1nF | 2MHz |

*Figure 5.20. Using a Schmitt trigger i.c. as an oscillator. Any inverting Schmitt circuit can be used. The table shows very approximately the values of frequency obtained using a 390 ohm resistor and various capacitor values*

voltage must be +5.0 V ± 0.25 V, and large operating currents are needed if more than a few TTL i.c.s are in use. Each group of five TTL i.c.s should have its power supply lines decoupled to avoid transmitting pulses back through the supply lines (even when a stabilised supply is used). The operating pulses should have fast rise and fall times, of the order of 50 ns or less, since many TTL circuits can oscillate if they are linearly biased, as can happen momentarily during a slow rising or falling input. A useful tip is to use Schmitt trigger gates at the input to each circuit, so ensuring a fast rising and falling pulse from any input. These circuits can also be used as clock-pulse generators (*Figure 5.20*).

*Table 5.4* (p. 142) shows the pin arrangements for a number of TTL i.c.s.

Most of the TTL range are now available as low power Schottky types. These i.c.s make use of a component called the Schottky diode, which can be built in i.c. form, and which limits the voltage between the collector and the base of a transistor when connected between these points. The use of Schottky diodes makes it possible to design gates which need much smaller currents than conventional TTL i.c.s and which, because the transistors never saturate, can switch as fast, or even faster. The typical input current which has to be sunk to keep a low-power Schottky gate input at logic 0 is around 0.4 mA, only a quarter of the amount needed for a conventional stage. These Schottky i.c.s are generally distinguished by the use of the letters LS in the type number, such as 74LS00, 74LS132, etc.

## MOS circuits

Small-scale digital circuits, as well as large-scale microprocessor or calculator chips, can be made using MOS techniques, making use of n-channel, p-channel, or both. One very useful series, the 4000 series, uses both p and n channel FETs, and is known as CMOS (Complementary MOS). Typical input and output circuits are shown in *Figure*

*Figure 5.21. Typical CMOS input and output circuits — the example is of an inverter. In practice, protection diodes are built in at the gate inputs to prevent excessive voltages from damaging the gates, so that nothing short of a visible spark at the input is likely to cause harm*

*5.21.* Since the input is always to the gate of a FET, the input resistance is always very high, so that virtually no d.c. will flow in the input circuit, except through protective diodes. This feature makes the fanout capability for low frequency signals very large, though at higher operating speeds the fanout is limited by the need for the output currents to charge and discharge the stray capacitances at the gate inputs. Because of the high input resistance, the input gates are easily damaged by electrostatic charges. When the i.c. is in circuit, load resistors, along with built-in protection diodes, guard against overloads, but when the i.c. is not in a circuit protection against electrostatic damage can be assured only while all the pins are shorted together. For storage, the pins are often embedded in conductive foam, or the i.c. contained in a conductive plastic case. Particular care has to be taken when CMOS i.c.s are connected into circuit. In 'normal' domestic surroundings, the following precautions are adequate.

(1) The remainder of the circuit should be complete before the MOS i.c.s are added.

(2) Plugging into holders is less hazardous than soldering.

(3) The negative line of the circuit should be earthed when the i.c.s are plugged in.

(4) If soldering is used, the soldering iron must have an earthed bit.

(5) No input pins must ever be left unconnected in a circuit.

(6) The pins of the i.c. should not be handled.

For industrial conditions where low humidity and large insulating areas can present unusual electrostatic problems, the manufacturers' guides should be consulted.

The high input resistance of CMOS i.c.s makes some circuits, particularly oscillator circuits, much easier to implement, and the very low current consumption makes the use of battery operation possible for large circuits. The maximum operating voltage is +15 V, and satisfactory operation is possible at only 3 V; a safe range of operating voltages is 4 V to 12 V. The delay timer for a typical gate is greater than that of a comparable TTL i.c., but for many applications this is not important. Clock pulses with rise or fall times exceeding 5 $\mu$s should not be used.

*Table 5.5* (p. 163) shows the pin arrangement of a range of CMOS i.c.s.

### The microprocessor

The microprocessor is a universal logic circuit whose output or outputs are under the control of a program stored in a memory. Any problem in combinational or sequential logic which could be solved by a set of logic i.c.s (a *hardware* solution) can be solved by a microprocessor, given a suitable program (a *software* solution). At the time of writing, three distinct types of microprocessor (disregarding the bit-slice types) are available:

(a) The 'dedicated' microprocessor, which is often designed with a built-in program for one particular type of use (such as games machines, petrol pumps, door-chimes).

(b) The general-purpose microprocessor which is, however, designed particularly for industrial control use (such as the F8, INS8060).

(c) The microcomputer type, intended to form the basis of miniature computers or of 'intelligent' machines (such as the Z80, 6800).

Within the space of this book it is impossible to discuss details of microprocessor operation, but a few general hints which are not always available from other sources may be helpful.

The microprocessor is generally a 40-pin package, but the connections which are needed vary very considerably from one design to another. A few types, in categories (a) and (b) may include a clock pulse generator, but most do not, so that an external clock signal, sometimes two-phase or even four-phase, may be needed. A suitable i.c. is usually available from the manufacturer of the microprocessor.

Address pins, usually 16 for the type (c) microprocessors, are used for output signals which select external circuits, memory mainly, but also input or output circuits, displays and other external circuits. Using 16 lines, 65 536 different code numbers can be specified, since a binary number consisting of sixteen 1s is $2^{16}$, hexadecimal FFFF, decimal 65 536. Each set of connections which is selected by an address number can be a stored binary number (8 bits for most of the smaller microprocessors, but 16-bit micros are now available) from a memory i.c. In general, only a very large and complex system would use all the address lines for memory addressing.

The data pins, usually eight for the smaller microprocessors, are used for input or output of binary signals in the form of 8-bit groups or bytes. These data pins are bidirectional, arranged so that control signals sent out by the microprocessor will designate the data pins as inputs, outputs, or high impedance. If the data pins can be placed in a high impedance, or isolated state, the data pins are said to have three-state control.

The remaining pins, apart from power supplies, are inputs which are used to control the action of the microprocessor (reset, wait, interrupt) and outputs which are used by the microprocessor to control other devices (memory request, read/write, bus request).

The microprocessor itself, unless it is equipped with internal memory circuits, is incapable of any useful functions, so that the arrangement of other i.c.s connected to the microprocessor is important; such an arrangement constitutes a system. A system will consist of the microprocessor or CPU (central processing unit), various memories which may be read/write (known as RAM) or the read-only (ROM) types. The Read-Only Memory is used for system instructions which must be available whenever the microprocessor system is switched on. The read/write memories, once known as Random Access Memories (hence RAM) in the days when other types, operated in sequence, were used, are volatile, meaning that the stored information is lost when the system is switched off. At switch-on, each RAM will contain a random arrangement of bits (garbage), but the ROM will still contain its program instructions, since ROM is non-volatile.

The remaining units of the system consist of peripheral devices such as PIAs (Peripheral Interface Adaptors) whose task is to accept or deliver 8-bit bytes to or from the data pins. It is the range of peripherals of this type which make the microprocessor capable of all its applications, permitting keyboard insertion of programs and data, output signals which will operate displays or switch other devices, such as thyristors, the recording of programs on tape or magnetic discs (floppies), the reading of programs or data from such sources, the operation of teleprinters and visual display units (VDUs) and so on. Each PIA is controlled either by the address line, as if it were part of the memory (memory mapped use) or by special signals from pins on the CPU i.c.

Because the data lines are bidirectional, all microprocessor operations must be in sequence. A program for displaying a number selected by a key on a keyboard for example, must instruct the device to read data into the CPU whenever any key is pressed, to identify the key which is pressed, generate a code to operate the display, and output that code to the correct display. Each instruction will take several steps of program, depending on the type of microprocessor which is used, since the PIA must be separately instructed by a code signal whether it is to function as an input or as an output at each part of the sequence.

For more detailed descriptions, the reader is referred to specific data books for the microprocessor which is to be used. In general, the more complex the system and the higher the programming language (the more like English, as distinct from number codes), the easier programming becomes. Programming of any length in machine code even for relatively simple applications is a very exhausting and frustrating business.

## Table 5.4. TTL PINOUTS

**7400**
**QUADRUPLE 2-INPUT**
**POSITIVE-NAND GATES**

**7401**
**QUADRUPLE 2-INPUT**
**POSITIVE-NAND GATES**
**WITH OPEN-COLLECTOR OUTPUTS**

**74H01**

**7402**
**QUADRUPLE 2-INPUT**
**POSITIVE-NOR GATES**

**7403**
**QUADRUPLE 2-INPUT**
**POSITIVE-NAND GATES**
**WITH OPEN-COLLECTOR OUTPUTS**

**7404**
**HEX INVERTERS**

**7405**
**HEX INVERTERS**
**WITH OPEN-COLLECTOR OUTPUTS**

**7406**
**HEX INVERTER BUFFERS/DRIVERS**
**WITH OPEN-COLLECTOR**
**HIGH-VOLTAGE OUTPUTS**

**7407**
**HEX BUFFERS/DRIVERS**
**WITH OPEN-COLLECTOR**
**HIGH-VOLTAGE OUTPUTS**

**7408**
**QUADRUPLE 2-INPUT**
**POSITIVE-AND GATES**

**7409**
**QUADRUPLE 2-INPUT**
**POSITIVE-AND GATES**
**WITH OPEN-COLLECTOR OUTPUTS**

**7410**
**TRIPLE 3-INPUT**
**POSITIVE-NAND GATES**

**74H11**
**TRIPLE 3-INPUT**
**POSITIVE-AND GATES**

**7412**
**TRIPLE 3-INPUT**
**POSITIVE-NAND GATES**
**WITH OPEN-COLLECTOR OUTPUTS**

**7413**
**DUAL 4-INPUT**
**POSITIVE-NAND**
**SCHMITT TRIGGERS**

**7414**
**HEX SCHMITT-TRIGGER**
**INVERTERS**

**74H15**
**TRIPLE 3-INPUT**
**POSITIVE-AND GATES**
**WITH OPEN-COLLECTOR OUTPUTS**

**7416**
**HEX INVERTER BUFFERS/DRIVERS**
**WITH OPEN-COLLECTOR**
**HIGH-VOLTAGE OUTPUTS**

**7417**
**HEX BUFFERS/DRIVERS**
**WITH OPEN-COLLECTOR**
**HIGH-VOLTAGE OUTPUTS**

**7420**
**DUAL 4-INPUT**
**POSITIVE-NAND GATES**

**74H21**
**DUAL 4-INPUT**
**POSITIVE-AND GATES**

**7422**
**DUAL 4-INPUT**
**POSITIVE-NAND GATES**
**WITH OPEN-COLLECTOR OUTPUTS**

**7423**
**EXPANDABLE DUAL 4-INPUT**
**POSITIVE-NOR GATES**
**WITH STROBE**

**7425**
**DUAL 4-INPUT**
**POSITIVE-NOR GATES**
**WITH STROBE**

**7426**
**QUADRUPLE 2-INPUT**
**HIGH-VOLTAGE INTERFACE**
**POSITIVE-NAND GATES**

**7427**
**TRIPLE 3-INPUT**
**POSITIVE-NOR GATES**

**7428**
**QUADRUPLE 2-INPUT**
**POSITIVE-NOR BUFFERS**

### 7430
### 8-INPUT
### POSITIVE-NAND GATES

### 7432
### QUADRUPLE 2-INPUT
### POSITIVE-OR GATES

### 7433
### QUADRUPLE 2-INPUT
### POSITIVE-NOR BUFFERS
### WITH OPEN-COLLECTOR OUTPUTS

### 7437
### QUADRUPLE 2-INPUT
### POSITIVE-NAND BUFFERS

### 7438
### QUADRUPLE 2-INPUT
### POSITIVE-NAND BUFFERS
### WITH OPEN-COLLECTOR OUTPUTS

### 7440
### DUAL 4-INPUT
### POSITIVE-NAND BUFFERS

### 7442A
### 4 LINE-TO-10-LINE DECODERS

### 7445
### BCD-TO-DECIMAL DECODER/DRIVER

### 7446
### BCD-TO-SEVEN-SEGMENT
### DECODERS/DRIVERS

### 7448
### BCD-TO-SEVEN-SEGMENT
### DECODERS/DRIVERS

### 74LS49
### BCD-TO-SEVEN-SEGMENT
### DECODERS/DRIVERS

### 7450
### DUAL 2-WIDE 2-INPUT
### AND-OR-INVERT GATES
### (ONE GATE EXPANDABLE)

### 7451
### AND-OR-INVERT GATES

MAKE NO EXTERNAL CONNECTION

### 74L51

### 74H52
### EXPANDABLE 4-WIDE
### AND-OR GATES

**7453**
**EXPANDABLE 4-WIDE**
**AND-OR-INVERT GATES**

**74H53**

**7454**
**4-WIDE**
**AND-OR-INVERT GATES**

**74H54**

**74L54**

**74H55**
**2-WIDE 4-INPUT**
**AND-OR-INVERT GATES**

**74L55**

**7460**
**DUAL 4-INPUT EXPANDERS**

**74H61**
**TRIPLE 3-INPUT**
**EXPANDERS**

**74H62**
**4-WIDE AND-OR EXPANDERS**

**74LS63**
**HEX CURRENT-SENSING**
**INTERFACE GATES**

**74S64**
**4-2-3-2 INPUT AND-OR-INVERT**
**GATES**

**7470**
**AND-GATED J-K POSITIVE-EDGE-**
**TRIGGERED FLIP-FLOPS**
**WITH PRESET AND CLEAR**

**74H71**
**AND-OR-GATED J-K MASTER-SLAVE**
**FLIP-FLOPS WITH PRESET**

**74L71**
**AND-GATED R-S MASTER-SLAVE**
**FLIP-FLOPS WITH PRESET**
**AND CLEAR**

### 7472
**AND-GATED J-K MASTER-SLAVE FLIP-FLOPS WITH PRESET AND CLEAR**

### 7473
**DUAL J-K FLIP-FLOPS WITH CLEAR**

### 7474
**DUAL D-TYPE POSITIVE-EDGE-TRIGGERED FLIP-FLOPS WITH PRESET AND CLEAR**

### 7475
**4-BIT BISTABLE LATCHES**

### 7476
**DUAL J-K FLIP-FLOPS WITH PRESET AND CLEAR**

### 74H78
**DUAL J-K FLIP-FLOPS WITH PRESET, COMMON CLEAR, AND COMMON CLOCK**

### 74L78

### 7480
**GATED FULL ADDERS**

### 7481
**16-BIT RANDOM-ACCESS MEMORIES**

### 7482
**2-BIT BINARY FULL ADDERS**

### 74 83
**4-BIT BINARY FULL ADDERS WITH FAST CARRY**

### 7484
**16-BIT RANDOM-ACCESS MEMORIES**

### 7485
**4-BIT MAGNITUDE COMPARATORS**

### 7486
**QUADRUPLE 2-INPUT EXCLUSIVE-OR GATES**

### 74L86

## 74H87
### 4-BIT TRUE/COMPLEMENT, ZERO/ONE ELEMENTS

## 7488A
### 256-BIT READ-ONLY MEMORIES

## 7489
### 64-BIT READ/WRITE MEMORIES

## 7490A
### DECADE COUNTERS

## 7491A
### 8-BIT SHIFT REGISTERS

## 7492
### DIVIDE-BY-TWELVE COUNTERS

## 7493A
### 4-BIT BINARY COUNTERS

## 7494
### 4-BIT SHIFT REGISTERS

## 7495A
### 4-BIT SHIFT REGISTERS

## 7496
### 5-BIT SHIFT REGISTERS

## 7497
### SYNCHRONOUS 6-BIT BINARY RATE MULTIPLIERS

## 74L98
### 4-BIT DATA SELECTOR/STORAGE REGISTERS

## 74L99
### 4-BIT BIDIRECTIONAL UNIVERSAL SHIFT REGISTERS

## 74100
### 8-BIT BISTABLE LATCHES

### 74H101
AND-OR-GATED J-K NEGATIVE-EDGE-TRIGGERED FLIP-FLOPS WITH PRESET

### 74H102
AND-GATED J-K NEGATIVE-EDGE-TRIGGERED FLIP-FLOPS WITH PRESET AND CLEAR

### 74H103
DUAL J-K NEGATIVE-EDGE-TRIGGERED FLOPS WITH CLEAR

### 74H106
DUAL J-K NEGATIVE-EDGE-TRIGGERED FLIP-FLOPS WITH PRESET AND CLEAR

### 74107
DUAL J-K FLIP-FLOPS WITH CLEAR

### 74H108
DUAL J-K NEGATIVE-EDGE-TRIGGERED FLIP-FLOPS WITH PRESET, COMMON CLEAR, AND COMMON CLOCK

### 74109
DUAL J-$\overline{K}$ POSITIVE-EDGE-TRIGGERED FLIP-FLOPS WITH PRESET AND CLEAR

### 74110
AND-GATED J-K MASTER-SLAVE FLIP-FLOPS WITH DATA LOCKOUT

### 74111
DUAL J-K MASTER-SLAVE FLIP-FLOPS WITH DATA LOCKOUT

### 74LS112A
DUAL J-K NEGATIVE-EDGE-TRIGGERED FLIP-FLOPS WITH PRESET AND CLEAR

### 74LS113A
DUAL J-K NEGATIVE-EDGE-TRIGGERED FLIP-FLOPS WITH PRESET

### 74LS114A
DUAL J-K NEGATIVE-EDGE-TRIGGERED FLIP-FLOPS WITH PRESET, COMMON CLEAR, AND COMMON CLOCK

### 74116
DUAL 4-BIT LATCHES

### 74120
DUAL PULSE SYNCHRONIZERS/DRIVERS

### 74121
MONOSTABLE MULTIVIBRATORS

### 74122
RETRIGGERABLE MONOSTABLE
MULTIVIBRATORS WITH CLEAR

### 74123
DUAL RETRIGGERABLE MONOSTABLE
MULTIVIBRATORS WITH CLEAR

### 74LS124
DUAL VOLTAGE-CONTROLLED
OSCILLATORS

### 74125
QUADRUPLE BUS BUFFER GATES
WITH THREE-STATE OUTPUTS

### 74126
QUADRUPLE BUS BUFFER GATES
WITH THREE-STATE OUTPUTS

### 74128
SN74128 ... 50-OHM LINE DRIVER

### 74132
QUADRUPLE 2-INPUT POSITIVE-
NAND SCHMITT TRIGGERS

### 74S133
13-INPUT POSITIVE-NAND GATES

### 74S134
12-INPUT POSITIVE-NAND GATES
WITH THREE-STATE OUTPUTS

### 74S135
QUAD EXCLUSIVE-OR/NOR GATES

### 74136
QUAD EXCLUSIVE-OR GATES

### 74LS138
3-TO-8 LINE DECODERS/
MULTIPLEXERS

### 74LS139
DUAL 2-TO-4 LINE DECODERS/
MULTIPLEXERS

### 74S140
DUAL 4-INPUT POSITIVE-NAND
50-OHM LINE DRIVERS

### 74141
**BCD-TO-DECIMAL DECODER/DRIVER**

### 74142
**COUNTER/LATCH/DECODER/DRIVER**

### 74143 74144
**COUNTERS/LATCHES/DECODERS/DRIVERS**

### 74145
**BCD-TO-DECIMAL DECODERS/DRIVERS FOR LAMPS, RELAYS, MOS**

### 74147
**10-LINE DECIMAL TO 4-LINE BCD PRIORITY ENCODERS**

### 74148
**8-LINE-TO-3-LINE OCTAL PRIORITY ENCODERS**

### 74150
**1-OF-16 DATA SELECTORS/MULTIPLEXERS**

### 74151A
**1-OF-8 DATA SELECTORS/MULTIPLEXERS**

### 74153
**DUAL 4-LINE TO 1-LINE DATA SELECTORS/MULTIPLEXERS**

### 74154
**4-LINE TO 16-LINE DECODERS/DEMULTIPLEXERS**

### 74155
**DECODERS/DEMULTIPLEXERS**

### 74157
**QUAD 2-TO 1-LINE DATA SELECTORS/MULTIPLEXERS**

### 74159
**4- TO 16-LINE DECODERS/DEMULTIPLEXERS**

### 74160 74161 74162 74163
**SYNCHRONOUS 4-BIT COUNTERS**

**74164**
**8-BIT PARALLEL OUTPUT SERIAL SHIFT REGISTERS**

**74165**
**PARALLEL-LOAD 8-BIT SHIFT REGISTERS WITH COMPLEMENTARY OUTPUTS**

**74166**
**8-BIT SHIFT REGISTERS**

**74167**
**SYNCHRONOUS DECADE RATE MULTIPLIERS**

**74S168    74S169**
**4-BIT UP/DOWN SYNCHRONOUS COUNTERS**

**74170**
**4-BY-4 REGISTER FILES**

**74172**
**16-BIT REGISTER FILE**

**74173**
**4-BIT D-TYPE REGISTERS**

**74174**
**HEX D-TYPE FLIP-FLOPS**

**74175**
**QUAD D-TYPE FLIP-FLOPS**

### 74176   74177
### PRESETABLE COUNTERS/LATCHES

### 74178
### 4-BIT UNIVERSAL SHIFT REGISTERS

### 74179
### 4-BIT UNIVERSAL SHIFT REGISTERS

### 74180
### 9-BIT ODD/EVEN PARITY
### GENERATORS/CHECKERS

### 74181
### ARITHMETIC LOGIC UNITS/FUNCTION
### GENERATORS

### 74182
### LOOK-AHEAD CARRY GENERATORS

### 74LS183
### DUAL CARRY-SAVE FULL ADDERS

### 74184   74185A
### CODE CONVERTERS

### 74186
### 512-BIT PROGRAMABLE READ-ONLY MEMORIES

### 74187
### 1024-BIT READ-ONLY MEMORIES

## 74188A
### 256-BIT PROGRAMMABLE READ-ONLY MEMORIES

## 74S189
### 64-BIT RANDOM-ACCESS MEMORIES

## 74190  74191
### SYNCHRONOUS UP/DOWN COUNTERS

## 74192  74193
### SYNCHRONOUS UP/DOWN DUAL CLOCK COUNTERS

## 74194
### 4-BIT BIDIRECTIONAL UNIVERSAL SHIFT REGISTERS

## 74195
### 4-BIT PARALLEL-ACCESS SHIFT REGISTERS

## 74196  74197
### PRESETABLE COUNTERS/LATCHES

## 74198
### 8-BIT BIDIRECTIONAL UNIVERSAL SHIFT REGISTERS

## 74199
### 8-BIT BIDIRECTIONAL UNIVERSAL SHIFT REGISTERS

## 74LS200A
### 256-BIT RANDOM-ACCESS MEMORIES

**74S201
256-BIT RANDOM-ACCESS MEMORIES**

**74LS202
256-BIT READ/WRITE MEMORIES WITH
POWER DOWN**

**74LS207
RANDOM-ACCESS MEMORIES**

**74LS208
RANDOM-ACCESS MEMORIES**

**74LS214    74LS215
RANDOM-ACCESS MEMORIES**

**74221
DUAL MONOSTABLE MULTIVIBRATORS**

**74S225
ASYNCHRONOUS FIRST IN, FIRST OUT MEMORIES**

**74S226
4-BIT PARALLEL LATCHED BUS TRANSCEIVERS**

**74LS240
OCTAL BUFFERS/LINE DRIVERS/LINE
RECEIVERS**

**74LS241
OCTAL BUFFERS/LINE DRIVERS/LINE
RECEIVERS**

**74LS242**
**QUADRUPLE BUS TRANSCEIVERS**

**74LS243**
**QUADRUPLE BUS TRANCEIVERS**

**74LS244**
**OCTAL BUFFERS/LINE DRIVERS/LINE RECEIVERS**

**74LS245**
**OCTAL BUS TRANCEIVERS**

**74246  74247**
**BCD-TO-SEVEN-SEGMENT**
**DECODERS/DRIVERS**

**74248  74249**
**BCD-TO-SEVEN-SEGMENT**
**DECODERS/DRIVERS**

**74251**
**DATA SELECTORS/**
**MULTIPLEXERS**

**74LS253**
**DUAL DATA SELECTORS/MULTIPLEXERS**

**74LS257A**
**QUAD DATA SELECTORS/MULTIPLEXERS**

**74LS258A**
**QUAD DATA SELECTORS/MULTIPLEXERS**

**74259**
**EIGHT-BIT ADDRESSABLE LATCHES**

**74S260**
**DUAL 5-INPUT POSITIVE NOR GATES**

**74265**
**QUAD COMPLEMENTARY-OUTPUT ELEMENTS**

**74S270**
**2048-BIT READ-ONLY MEMORIES**

**74273**
**OCTAL D-TYPE FLIP-FLOPS**

**74LS275**
**7-BIT SLICE WALLACE TREES**

**74LS261**
**2-BIT BY 4-BIT PARALLEL BINARY MULTIPLIERS**

**74LS266**
**QUAD 2-INPUT EXCLUSIVE-NOR GATES WITH OPEN-COLLECTOR OUTPUTS**

**74S271**
**2048-BIT READ-ONLY MEMORIES**

**74S274**
**4-BIT BY 4-BIT BINARY MULTIPLIERS**

**74276**
**QUAD J-$\overline{\text{K}}$ FLIP-FLOPS**

**74278**
**4-BIT CASCADEABLE PRIORITY REGISTERS**

**74279**
**QUAD S-R LATCHES**

**74LS280**
**9-BIT ODD/EVEN PARITY GENERATORS/CHECKERS**

**74S281**
**4-BIT PARALLEL BINARY ACCUMULATORS**

**74283**
**4-BIT BINARY FULL ADDERS**

**74284**
**4-BIT-BY-4-BIT PARALLEL BINARY MULTIPLIERS USED WITH '285**

**74285**
**4-BIT-BY-4-BIT PARALLEL BINARY MULTIPLIERS USED WITH '284**

**74S287**
**1024-BIT PROGRAMMABLE READ-ONLY MEMORIES**

**74S288**
**256-BIT PROGRAMMABLE READ-ONLY MEMORIES**

**74S289**
**64-BIT RANDOM-ACCESS MEMORIES**

**74290**
**DECADE COUNTERS**

**74293**
**4-BIT BINARY COUNTERS**

**74LS295B**
**4-BIT BIDIRECTIONAL UNIVERSAL SHIFT REGISTERS**

**74298**
**QUAD 2-INPUT MULTIPLEXERS WITH STORAGE**

### 74LS299
### 8-BIT BIDIRECTIONAL UNIVERSAL SHIFT/STORAGE REGISTERS

### 74LS300A
### 256-BIT READ/WRITE MEMORIES

### 74S301
### 256-BIT RANDOM ACCESS MEMORIES

### 74LS302
### 256-BIT READ/WRITE MEMORIES

### 74LS314    74LS315
### 1024-BIT RANDOM-ACCESS MEMORIES

### 74LS323
### 8-BIT BIDIRECTIONAL UNIVERSAL SHIFT/STORAGE REGISTERS

### 74LS324
### VOLTAGE-CONTROLLED OSCILLATORS

### 74LS325
### DUAL VOLTAGE-CONTROLLED OSCILLATORS

### 74LS326
### DUAL VOLTAGE-CONTROLLED OSCILLATORS

### 74LS327
### DUAL VOLTAGE-CONTROLLED OSCILLATORS

### 74LS348
### 8-LINE-TO-3-LINE PRIORITY ENCODERS

### 74351
### DUAL 8-LINE-TO-1-LINE DATA SELECTOR/MULTIPLEXER

### 74LS352
### DUAL 4-LINE-TO-LINE DATA SELECTORS/MULTIPLEXERS

**74LS353**
**DUAL 4-LINE-TO-1-LINE DATA**
**SELECTORS/MULTIPLEXERS**

**74LS362**
**FOUR-PHASE CLOCK GENERATOR/DRIVER**
**FOR TMS 9900 MICROPROCESSOR**

**74LS363**
**OCTAL D-TYPE LATCHES**

**74LS364**
**OCTAL D-TYPE FLIP-FLOPS**

**174365A**
**HEX BUS DRIVERS**

**74366A**
**HEX BUS DRIVERS**

**74367A**
**HEX BUS DRIVERS**

**74368A**
**HEX BUS DRIVERS**

**74S370**
**2048-BIT READ-ONLY MEMORIES**

**74S371**
**2048-BIT READ-ONLY MEMORIES**

### 74LS373
### OCTAL D-TYPE LATCHES

### 74LS374
### OCTAL D-TYPE FLIP-FLOPS

### 74LS375
### 4-BIT BISTABLE LATCHES

### 74376
### QUAD J-K̄ FLIP-FLOPS

### 74LS377
### OCTAL D-TYPE FLIP-FLOPS

### 74LS378
### HEX D-TYPE FLIP-FLOPS

### 74LS379
### QUAD D-TYPE FLIP-FLOPS

### 74S381
### ARITHMETIC LOGIC UNITS/FUNCTION
### GENERATORS

### 74LS386
### QUAD 2-INPUT EXCLUSIVE-OR
### GATES

### 74S387
### 1024-BIT PROGRAMMABLE
### READ-ONLY MEMORIES

### 74390
### DUAL DECADE COUNTERS

### 74393
### DUAL 4-BIT BINARY COUNTERS

## 74LS395A
### 4-BIT UNIVERSAL SHIFT REGISTERS

## 74LS398
### QUAD 2-INPUT MULTIPLEXERS WITH STORAGE

## 74LS399
### QUAD 2-INPUT MULTIPLEXERS WITH STORAGE

## 74S412
### MULTI-MODE BUFFERED 8-BIT LATCHES

## 74LS424
### TWO-PHASE CLOCK GENERATOR/ DRIVER FOR 8080A

## 74425
### QUAD GATES

## 74426
### QUAD GATES

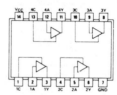

## 74S428    74S438
### SYSTEM CONTROLLER FOR 8080A

## 74S470    74S471
### PROGRAMMABLE READ-ONLY MEMORIES

## 74S472    74S473
### PROGRAMMABLE READ-ONLY MEMORIES

## 74S474    74S475
### PROGRAMMABLE READ-ONLY MEMORIES

**74S481**
**4-BIT SLICE PROCESSOR ELEMENTS**

**74S482**
**4-BIT-SLICE EXPANDABLE CONTROL ELEMENTS**

**74490**
**DUAL DECADE COUNTERS**

**74LS670**
**4-BY-4 REGISTER FILES**

## Table 5.5. CMOS PINOUTS

CD4000
Dual 3-Input NOR Gate
Plus Inverter

CD4001
Quad 2-Input NOR Gate

CD4002
Quad 4-Input NOR Gate

CD4006
8-Stage Static Shift
Register

CD4007
Dual Complementary Pair
Plus Inverter

CD4008
4-Bit Full Adder with
Parallel Carry Out

## CD4009
### Hex Buffer/Converter
### Inverting Type

$G = \overline{A}$
$H = \overline{B}$
$I = \overline{C}$
$J = \overline{D}$
$K = \overline{E}$
$L = \overline{F}$

$V_{CC} = 1$
$V_{SS} = 8$   $V_{DD} = 16$
NC = 13

## CD4010
### Hex Buffer/Converter
### Non-Inverting Type

$G = A$
$H = B$
$I = C$
$J = D$
$K = E$
$L = F$

$V_{CC} = 1$
$V_{SS} = 8$   $V_{DD} = 16$
NC = 13

## CD4011
### Quad 2-Input NAND Gate

$J = \overline{A \cdot B}$
$K = \overline{C \cdot D}$
$L = \overline{E \cdot F}$
$M = \overline{G \cdot H}$

## CD4012
### Dual 4-Input NAND Gate

$J = \overline{A \cdot B \cdot C \cdot D}$
$K = \overline{E \cdot F \cdot G \cdot H}$

## CD4013
### Dual "D" Flip-Flop with
### Set/Reset Capability

## CD4014
### 8-Stage Synchronous Shift
### Register with Parallel or
### Serial Input/Serial Output

## CD4015
### Dual 4-Stage Static Shift Register
### with Serial Input/Parallel Output

## CD4016
### Quad Bilateral Switch

## CD4017
### Decade Counter/Divider with
### 10 Decoded Decimal Outputs

## CD4018

### Presettable Divide-by-"N" Counter Fixed or Programmable

## CD4019

### Quad AND/OR Select Gate

$D_4 = (A_4 \cdot K_a) + (B_4 \cdot K_b)$

## CD4020

### 14-Stage Binary Ripple Counter

## CD4021

### 8-Stage Static Shift Register Asynchronous Parallel or Synchronous Serial Input/Serial Output

## CD4022

### Divide-by-8 Counter/Divider with 8 Decoded Decimal Outputs

$V_{DD} = 16$
$V_{SS} = 8$

## CD4023

### Triple 3-Input NAND Gate

$L = G \cdot H \cdot I$

$K = \overline{D \cdot E \cdot F}$

$J = \overline{A \cdot B \cdot C}$

## CD4024

### 7-Stage Ripple-Carry Binary Counter/Divider

NC = 8, 10, 13

## CD4025

### Triple 3-Input NOR Gate

$L = \overline{G + H + I}$

$K = \overline{D + E + F}$

$J = \overline{A + B + C}$

## CD4026

### Decade Counter/Divider with 7-Segment Display Outputs and Display Enable

### CD4027
Dual J-K Master-Slave Flip-Flop with Set-Reset Capability

### CD4028
BCD-to-Decimal Decoder

### CD4029
Presettable Up/Down Counter, Binary or BCD-Decade

### CD4030
Quad Exclusive-OR Gate

$J = A \oplus B$   $L = E \oplus F$
$K = C \oplus D$   $M = G \oplus H$

### CD4031
64-Stage Static Shift Register

$V_{DD} = 16$
$V_{SS} = 8$
NC = 3, 4, 11, 12, 13, 14

### CD4032
Triple Serial Adder Positive Logic

$V_{SS} = 8$
$V_{DD} = 16$

### CD4033
Decade Counter/Divider with 7-Segment Display Outputs and Ripple Blanking

### CD4034
8-Stage Static Bidirectional Parallel/Serial Input/Output Bus Register

### CD4035
4-Stage Parallel In/Parallel Out Shift Register with J-K Serial Inputs and True/Complement Outputs

## CD4037
### Triple AND/OR Bi-Phase Pair

$V_{DD} = 14$
$V_{CC} = 1$
$V_{SS} = 7$

## CD4038
### Triple Serial Adder Negative Logic

$V_{SS} = 8$
$V_{DD} = 16$

## CD4040
### 12-Stage Ripple-Carry Binary Counter/Divider

## CD4041
### Quad True/Complement Buffer

$V_{SS} = 7$
$V_{DD} = 14$

## CD4042
### Quad Clocked "D" Latch

## CD4043
### Quad 3-State NOR R/S Latch

## CD4044
### Quad 3-State NAND R/S Latch

## CD4045
### 21-Stage Counter

$V_{DD} = 3$
$V_{SS} = 14$

4, 5, 6, 9, 10, 11, 12, 13 = NO CONNECTION

## CD4046
### Micropower Phase-Locked Loop

## CD4047
### Low-Power Monostable/Astable
### Multivibrator

## CD4048
### Multi-Function Expandable
### 8-Input Gate

$V_{SS} = 8$
$V_{DD} = 16$

## CD4049
### Hex Buffer/Converter
### Inverting Type

A $\frac{3}{}$ ▷ $\frac{2}{}$ $G = \bar{A}$

B $\frac{5}{}$ ▷ $\frac{4}{}$ $H = \bar{B}$

C $\frac{7}{}$ ▷ $\frac{6}{}$ $I = \bar{C}$

D $\frac{9}{}$ ▷ $\frac{10}{}$ $J = \bar{D}$

E $\frac{11}{}$ ▷ $\frac{12}{}$ $K = \bar{E}$

F $\frac{14}{}$ ▷ $\frac{15}{}$ $L = \bar{F}$

$V_{CC} \frac{1}{}$

$V_{SS} \frac{8}{}$

NC = 13
NC = 16

## CD4050
### Hex Buffer/Converter
### Non-Inverting Type

A $\frac{3}{}$ ▷ $\frac{2}{}$ $G = A$

B $\frac{5}{}$ ▷ $\frac{4}{}$ $H = B$

C $\frac{7}{}$ ▷ $\frac{6}{}$ $I = C$

D $\frac{9}{}$ ▷ $\frac{10}{}$ $J = D$

E $\frac{11}{}$ ▷ $\frac{12}{}$ $K = E$

F $\frac{14}{}$ ▷ $\frac{15}{}$ $L = F$

$V_{CC} \frac{1}{}$

$V_{SS} \frac{8}{}$

NC = 13
NC = 16

## CD4051
### Single 8-Channel Analog
### Multiplexer/Demultiplexer

## CD4052
**Differential 4-Channel Analog Multiplexer/Demultiplexer**

## CD4053
**Triple 2-Channel Multiplexer/Demultiplexer**

## CD4054
**4-Segment Liquid-Crystal Display Driver**

## CD4055
**BCD-to-7-Segment Decoder/Driver with "Display-Frequency" Output Liquid-Crystal Display Driver**

## CD4056
**BCD-to-7-Segment Decoder/Driver with Strobed-Latch Function Liquid-Crystal Display Driver**

## CD4057
### 4-Bit Arithmetic Logic Unit

## CD4059
### Programmable Divide-by-"N" Counter

## CD4060
### 14-Stage Ripple-Carry Binary Counter/Divider and Oscillator

## CD4062
### 200-Stage Dynamic Shift Register

## CD4063
### 4-Bit Magnitude Comparator

## CD4066
### Quad Bilateral Switch

### CD4067
16-Channel
Multiplexer/Demultiplexer

$V_{DD}$ = 24
$V_{SS}$ = 12

### CD4068
8-Input NAND/AND Gate

J = A·B·C·D·E·F·G·H
K = A·B·C·D·E·F·G·H

$V_{DD}$ = 14

$V_{SS}$ = 7    6,8 = NO CONNECTION

### CD4069
Hex Inverter

A $\frac{1}{}$ ▷o $\frac{2}{}$ G = $\bar{A}$

B $\frac{3}{}$ ▷o $\frac{4}{}$ H = $\bar{B}$

C $\frac{5}{}$ ▷o $\frac{6}{}$ I = $\bar{C}$

D $\frac{9}{}$ ▷o $\frac{8}{}$ J = $\bar{D}$

E $\frac{11}{}$ ▷o $\frac{10}{}$ K = $\bar{E}$

F $\frac{13}{}$ ▷o $\frac{12}{}$ L = $\bar{F}$

$V_{DD}$ = 14
$V_{SS}$ = 7

### CD4070
Quad Exclusive-OR Gate

J = A⊕B    M = G⊕H
K = C⊕D    L = E⊕F

$V_{SS}$ = 7
$V_{DD}$ = 14

### CD4071
Quad 2-Input OR Gate

### CD4072
Dual 4-Input OR Gate

### CD4073
Triple 3-Input AND Gate

### CD4075
Triple 3-Input OR Gate

### CD4076
4-Bit D-Type Register

$V_{SS}$ = 8
$V_{DD}$ = 16

## CD4077
### Quad Exclusive-NOR Gate

$\bar{J} = A \oplus B$    $\bar{M} = G \oplus H$
$\bar{K} = C \oplus D$    $\bar{L} = E \oplus F$

## CD4078
### 8-Input NOR/OR Gate

$J = \overline{A+B+C+D+E+F+G+H}$
$K = A+B+C+D+E+F+G+H$
$6, 8 = NC$

$V_{DD} = 14$

$V_{SS} = 7$

## CD4081
### Quad 2-Input AND Gate

## CD4082
### Dual 4-Input AND Gate

## CD4085
### Dual 2-Wide, 2-Input AND-OR-INVERT (AOI) Gate

$E = \overline{INHIBIT + AB + CD}$
LOGIC 1 ≡ HIGH
LOGIC 0 ≡ LOW

$V_{DD} = 14$
$V_{SS} = 7$

## CD4086
### Expandable 4-Wide, 2-Input AND-OR-INVERT (AOI) Gate

LOGIC 1 ≡ HIGH
LOGIC 0 ≡ LOW

$V_{DD} = 14$
$V_{SS} = 7$
$NC = 4$

$J = \overline{INH + ENABLE + AB + CD + EF + GH}$

## CD4089
### Binary Rate Multiplier

$V_{DD} = 16$
$V_{SS} = 8$

CD4093
Quad 2-Input NAND
Schmitt Trigger

CD4094
8-Stage Shift-and-Store
Bus Register

(TERMINALS 4, 5, 6, 7, 14, 13, 12, 11, RESPECTIVELY)

CD4095
Gated J-K Master-Slave
Flip-Flop, Non-Inverting
Inputs

CD4096
Gated J-K Master-Slave
Flip-Flop, Inverting and
Non-Inverting Inputs

CD22100
4-by-4 Crosspoint Switch
with Control Memory

## CD22101   CD22102
### 4-by-4-by-2 Crosspoint Switch with Control Memory

## CD40100
### 32-Stage Static Left/Right Shift Register

## CD40101
### 9-Bit Parity Generator/Checker

## CD40102 2-Decade BCD
## CD40103 8-Bit Binary
### 8-Stage Presettable Synchronous Down Counter

## CD40105
### FIFO Register
### 4-Bits Wide by 16-Bits Long

## CD40106
### Hex Schmitt Trigger

## CD40107
### Dual 2-Input NAND Buffer/Driver

## CD40108
### 4-by-4 Multiport Register

CD40109

**Quad Low-to-High
Voltage Level Shifter**

CD40110

**Decade Up-Down Counter/
Decoder/Latch/Driver**

CD40114

**64-Bit Random-Access Memory**

## CD40115
### 8-Bit Bidirectional
### CMOS/TTL Level Converter

DISABLE
13

CMOS
IN(OUT)

LEVEL SHIFTER

TTL OUT(IN)

8

8

2-9

14-21

ENABLE
10

LEVEL SHIFTER

$V_{DD} = 1$
$V_{CC} = 22$
$V_{SS} = 11$

## CD40147
### 10-Line-to-4-Line
### BCD Priority Encoder

9

PRIORITY SELECT

BCD ENCODER

D2 3
C2 2
B2 1
A2 0

0

CD40160 Decade with Asynchronous Clear
CD40161 Binary with Asynchronous Clear
CD40162 Decade with Synchronous Clear
CD40163 Binary with Synchronous Clear
**Synchronous 4-Bit Counter**

PE 7
TE 10
CLEAR 1
LOAD 9
CLOCK 2
P1 3
P2 4
P3 5
P4 6

14 Q1
13 Q2
12 Q3
11 Q4
15 CARRY OUT

$V_{DD} = 16$
$V_{SS} = 8$

## CD40174
### Hex "D" Type Flip-Flop

DI 3      F/F1      2 Q1
D2 4      F/F2      5 Q2
D3 6      F/F3      7 Q3
D4 11     F/F4      10 Q4
D5 13     F/F5      12 Q5
D6 14     F/F6      15 Q6
CLOCK 9
CLEAR 1

$V_{SS} = 8$
$V_{DD} = 16$

## CD4097
### Differential 8-Channel
### Multiplexer/Demultiplexer

INH

3

1 OF 8 DECODERS

0

X
IN/OUT

7

X
OUT/IN

0

Y
IN/OUT

7

Y
OUT/IN

$V_{DD} = 24$
$V_{SS} = 12$

## CD4098
### Dual Monostable
### Multivibrator

$C_{X1}$   $R_{X1}$
$V_{DD}$

1   2 $R_X C_X (1)$

+TR 4
-TR 5
RESET 3

MONO₁

6 Q1
7 $\overline{Q}_1$

+TR 12
-TR 11
RESET 13

MONO₂

10 Q2
9 $\overline{Q}_2$

15   14 $R_X C_X (2)$

$V_{DD} = 16$
$V_{SS} = 8$

$C_{X2}$   $R_{X2}$   $V_{DD}$

## CD4099
### 8-Bit Addressable Latch

WRITE DISABLE — 4
DATA — 3

8 LATCHES

A0 — 5
A1 — 6
A2 — 7
DECODER

RESET — 2

9 — Q0
10 — Q1
11 — Q2
12 — Q3
13 — Q4
14 — Q5
15 — Q6
1 — Q7

$V_{DD} = 16$
$V_{SS} = 8$

## CD4502
### Strobed Hex Inverter/Buffer

THREE-STATE
OUTPUT 4
DISABLE
INHIBIT 12
D1 3
D2 6
D3 1
D4 10
D5 13
D6 15

5 Q1
7 Q2
2 Q3
9 Q4
11 Q5
14 Q6

$V_{DD} = 16$
$V_{SS} = 8$

## CD4508
### Dual 4-Bit Latch

OUTPUT DISABLE
D0A
D1A
D2A
D3A
STROBE
RESET

4-BIT LATCH   3-STATE OUTPUTS

Q0A
Q1A
Q2A
Q3A

OUTPUT DISABLE
D0B
D1B
D2B
D3B
STROBE
RESET

4-BIT LATCH   3-STATE OUTPUTS

Q0B
Q1B
Q2B
Q3B

## CD4510
### BCD Presettable Up/Down Counter

PRESET ENABLE — 1

P1 — 4
P2 — 12
P3 — 13
P4 — 3

CLOCK — 15
UP/DOWN — 10
CARRY IN — 5

RESET — 9

6 — Q1
11 — Q2
14 — Q3
2 — Q4

$V_{DD} = 16$
$V_{SS} = 8$

7 — CARRY OUT

## CD4511
### BCD-to-7-Segment Latch Decoder Driver

BCD INPUTS        7-SEGMENT OUTPUTS
LT
3

A — 7
B — 1
C — 2
D — 6

LATCH   DECODER   DRIVER

13 a
12 b
11 c
10 d
9 e
15 f
14 g

5
LE/STROBE   4   BL

$V_{SS} = 8$
$V_{DD} = 16$

## CD4512
### 8-Channel Data Selector

3-STATE DISABLE
INHIBIT
10   15

CHANNEL INPUTS
D0 — 1
D1 — 2
D2 — 3
D3 — 4
D4 — 5
D5 — 6
D6 — 7
D7 — 9

SELECT CONTROL
A — 11
B — 12
C — 13

14 SELECT OUTPUT

$V_{DD} = 16$
$V_{SS} = 8$

CD4512B

CD4514 Output "High" on Select
CD4515 Output "Low" on Select

**4-Bit Latch/4-to-16
Line Decoder**

CD4516

**Binary Presettable Up/Down
Counter**

CD4517

**Dual 64-Bit Shift Register**

CD4518 BCD
CD4520 Binary

**Dual Up Counter**

CD4527

**BCD Rate Multiplier**

CD4532

**8-Bit Priority Encoder**

CD4536

**Programmable Timer**

### CD4555

Dual Binary-to-1-of-4
Decoder/Demultiplexer
Output "High" on Select

### CD4556I

Dual Binary-to-1-of-4
Decoder/Demultiplexer
Output "Low" on Select

### CD4585

4-Bit Magnitude Comparator

### CD4724

8-Bit Addressable Latch

### CD40181

4-Bit Arithmetic Logic Unit

Active-Low Data      Active-High Data

### CD40182

Look-Ahead Carry Generator

### CD40192 BCD
### CD40193 Binary
Presettable Up/Down Counter
(Dual Clock with Reset)

## CD40208
### 4-by-4 Multiport Register

## CD40257
### Quad 2-Line-to-1-Line
### Data Selector/Multiplexer

# Appendix I

**Standard metric wire table**

| Diameter (mm) | Resistance (ohms per metre) | Current rating (mA) |
|---|---|---|
| 0.025 | 35.1 | 2.3 |
| 0.032 | 21.4 | 3.7 |
| 0.040 | 13.7 | 5.8 |
| 0.050 | 8.8 | 9.1 |
| 0.063 | 5.5 | 14.5 |
| 0.080 | 3.4 | 23.4 |
| 0.100 | 2.2 | 36.5 |
| 0.125 | 1.4 | 57.1 |
| 0.140 | 1.1 | 71.6 |
| 0.160 | 0.86 | 93.5 |
| 0.180 | 0.68 | 118.3 |
| 0.200 | 0.55 | 146.1 |
| 0.250 | 0.35 | 228.3 |
| 0.280 | 0.28 | 286.3 |
| 0.315 | 0.22 | 362.4 |
| 0.400 | 0.14 | 584.3 |
| 0.450 | 0.11 | 739.5 |
| 0.500 | 0.088 | 913.0 |
| 0.56 | 0.070 | 1.14 (amperes) |
| 0.63 | 0.055 | 1.45 |
| 0 71 | 0.043 | 1.84 |
| 0.75 | 0.039 | 2.05 |
| 0.80 | 0.034 | 2.34 |
| 0.085 | 0.030 | 2.64 |
| 0.90 | 0.027 | 2.96 |
| 0.95 | 0.024 | 3.30 |
| 1.00 | 0.022 | 3.65 |

The values of resistance per metre and of current rating have been rounded off. Only the smaller gauges are tabulated, representing the range of wire gauges which might be used in constructing r.f. and a.f. transformers.

# Appendix II

## Bibliography

Some of the most useful reference books in electronics are either out of print or difficult to obtain. They have been included in the list below because they can often be found in libraries, or in second-hand shops.

*Components: Understanding Electronic Components* (Sinclair) Fountain Press, 1972.

*Formulae and tables: Reference Data for Radio Engineers* (ITT).

*Audio & Radio: Radio Designers's Handbook* (Langford-Smith) Iliffe, 4th edn, 1967. A wealth of data, though often on valve circuits. Despite the age of the book, now out of print, it is still the most useful source book for audio work.

*Radio Amateurs Handbook* (ARRL). A mine of information of transmitting and receiving circuits. An excellent British counterpart is available, but the US publication contains more varied circuits, because the US amateur is not so restricted in his operations.

*GE Transistor Manual* (General Electric of USA). Even the early editions are extremely useful.

*Oscilloscopes: The Oscilloscope in Use* (Sinclair) Argus Books, 1976.

*Manufacturer's Databooks* by Texas, RCA, SGS–ATES, Motorola, National Semiconductor and Mullard contain detailed information on semiconductors, with many applications circuits.

*Microprocessors:* The textbooks by Dr. Adam Osborne are by far the most useful for anyone who already has some knowledge of micro-processors. For beginners, the last chapter of *Beginner's Guide to Digital Electronics* is a useful introduction.

A useful book on home computers and programming is *Microprocessors for Hobbyists* (Coles) Newnes Technical Books, 1979.

# Index